Lecture Notes in Social Networks

Lecture Notes in Social Networks (LNSN) comprises volumes covering the theory, foundations and applications of the new emerging multidisciplinary field of social networks analysis and mining. LNSN publishes peer- reviewed works (including monographs, edited works) in the analytical, technical as well as the organizational side of social computing, social networks, network sciences, graph theory, sociology, semantic web, web applications and analytics, information networks, theoretical physics, modeling, security, crisis and risk management, and other related disciplines. The volumes are guest-edited by experts in a specific domain. This series is indexed by DBLP. Springer and the Series Editors welcome book ideas from authors. Potential authors who wish to submit a book proposal should contact Annelies Kersbergen, Publishing Editor, Springer

e-mail: annelies.kersbergen@springer.com

David Knoke

Network Collective Action

Agent-Based Models of Pandemics, Riots, Social Movements, Insurrections and Insurgencies

Springer

David Knoke
Department of Sociology
University of Minnesota
Minneapolis, MN, USA

ISSN 2190-5428 ISSN 2190-5436 (electronic)
Lecture Notes in Social Networks
ISBN 978-3-031-86198-7 ISBN 978-3-031-86199-4 (eBook)
https://doi.org/10.1007/978-3-031-86199-4

This Springer imprint is published by the registered company Springer Nature Switzerland AG
The registered company address is: Gewerbestrasse 11, 6330 Cham, Switzerland

To my students.

Preface

This book is a product of the COVID-19 global pandemic. When the virus reached the University of Minnesota in the middle of Spring Semester 2020, faculty were told they had 1 week to convert their courses from in-classroom to online Zoom instruction. That experience was equivalent to learning to swim by being tossed into the deep end of a pool. Fortunately, I received a sabbatical for the following academic year and was spared from additional trauma.

I had proposed to review theories and empirical research on collective action by social networks in several substantive areas and attempt to anticipate future research directions. The term "agent-based model" appeared nowhere in the proposal. Two upheavals redirected my attention. As COVID-19 ravaged nation after nation, I became fascinated by epidemiological efforts to predict and suppress the spread of infections through interventions, such as social distancing, masking, and lockdowns. I learned that conventional compartmental models were complemented by computer microsimulations of dynamic person-to-person viral transmission. After President Trump lost his re-election bid, he fostered the violent January 6, 2021, attack on Capitol Hill to block Congress from certifying the Electoral College votes. Suspecting that agent-based modeling might be a useful method for exploring alternative insurrection scenarios, I set about learning how to write NetLogo computer codes. Over the next 3 years, my sabbatical project morphed into the half-dozen agent-based models of network collective action that comprise the core chapters of this book.

This book is not intended for advanced experts in ABM methods, although many examples of their work are reviewed. The simple models that I created illustrate one or two primary relations in collective action scenarios. Most result in collective outcomes of illness, injury, incarceration, or death of participants. They demonstrate that alternative outcomes may emerge if the initial conditions under which collective events occur are changed. In cases where two groups oppose one another, the victor is usually the side that mobilizes the most supporters. Not a great revelation, but a recurring experience ever since our human ancestors climbed down from the trees and began spreading across the plains of Africa.

Who should read this volume? Its target audience is social scientists who do not yet use ABMs but are curious to learn more. It's meant for people intrigued by the idea that complex social behaviors may be better understood as emergent phenomena resulting from interacting individuals within simulated environments. By running multiple experiments under varying conditions, researchers can investigate the outcomes of rare or unusual conditions. My hope is that more anthropologists, archaeologists, criminologists, economists, geographers, historians, linguists, political scientists, psychologists, sociologists, and other social scientists will be stimulated to add agent-based modeling to their theoretical and empirical tool kits.

Minneapolis, MN, USA David Knoke
January 17, 2025

Acknowledgements

I thank the University of Minnesota College of Liberal Arts for a 2020–2021 sabbatical that enabled me to start working on a project that eventuated in this book. I also thank the Undergraduate Research Opportunity Program for a small grant to Tiffany Wu and me to work on the ABM of recruitment to a social movement, as presented in the chapter "Recruiting."

Contents

About This Book

Chapter "Modeling" provides a brief overview of NetLogo, the software used in this book. The six featured ABMs in chapters "Infecting", "Crushing", "Rioting", "Recruiting", "Insurrecting", and "Insurging" are hosted on SpringerLink as electronic supplementary material. To use those models, readers should visit NetLogo at https://ccl.northwestern.edu/netlogo/ and download the NetLogo software. The programs are also available on the NetLogo Modeling Commons http://modelingcommons.org/account/login.

NetLogo is an open-source programming language and integrated development environment (IDE) for agent-based modeling. It was designed in 1999 by Uri Wilensky, Director of Northwestern University's Center for Connected Learning and Computer-Based Modeling (CCL). He intended it to be accessible to high-school students and professionals with limited computing skills in programming. NetLogo software can be downloaded for free, along with manuals, dictionaries, tutorials, and an online Modeling Commons, an archive for sharing and discussing more than a thousand user-contributed NetLogo ABMs spanning the natural sciences, art, philosophy, and social sciences (Wilensky and Rand 2015; Hamill and Gilbert 2016; Gilbert 2020). An indispensable document for learning how to create the ABMs presented in this book was the *NetLogo User Manual 6.4.0* (Wilensky 2023).

Theorizing

Keywords Collective action · Collective behavior · Collective decisions · Diffusion · Flocking · Theory · Thresholds

1 Introduction

On May 25, 2020, on a street outside a Minneapolis, Minnesota, convenience store, a white police officer pressed his knee for about 9 min into the neck of George Floyd until he died. Floyd, a Black man, had been arrested on suspicion of trying to pass a counterfeit $20 bill at the store. A passerby's video of the murder went viral, touching off protests and demonstrations around the world against police abuse and violence. Over the following nights in Minneapolis and St. Paul, crowds of angry protesters grew increasingly large and restless. Some people looted restaurants, ransacked retail stores, and set fire to buildings and cars, resulting in tens of millions of dollars in damages (Penrod and Sinner 2020). Jeering protestors gathered nightly outside the Minneapolis Police Department's 3rd Precinct, hurling rocks and fireworks. Officers defended their building by firing tear gas and rubber bullets, further escalating a volatile situation. On the night of May 28, the mob stormed the building. The mayor and police chef "eventually decided that having officers stand their ground wasn't worth the potential loss of life" (Caputo et al. 2020). As police in riot gear fled on foot, the mob breached the station and set it ablaze. After Minnesota's governor called out the National Guard, which took 2 days to mobilize, the violence subsided and protests and vigils turned largely peaceful.

The events that month in Minneapolis and St. Paul were a dramatic instance of collective action by groups of people assembling at the same time in one place. This book explores diverse examples of such behaviors using computer simulations called agent-based models to investigate alternative assumptions, dynamics, and outcomes of collective action events. This chapter defines collective action and collective behavior, presents a typology of events, and summarizes basic theories and models that seek to explain the causes and consequences of collective actions.

D. Knoke, *Network Collective Action*, Lecture Notes in Social Networks, https://doi.org/10.1007/978-3-031-86199-4_1

2 Collective Behavior and Collective Action

Among the earliest sociological concepts, collective behavior is closely related to collective action. Clark McPhail's intellectual history, *The Myth of the Madding Crowd* (1991), traced the idea of collective behavior back to Gustav LeBon's *The Psychology of Crowds* (1895), which argued that individuals who participate in crowds are transformed into a collective mind. Space limitations prevent a detailed recapitulation of the concept's many intertwined roots and branches across the following century. Instead this section concentrates on more recent similarities and distinctions between collective behavior and collective action. Keep in mind that definitions are neither true nor false, but only more or less useful in efforts to understand and explain some phenomenon of interest.

McPhail and Ronald Wohlstein (1983: 580–581) defined collective behavior as "two or more persons engaged in one or more behaviors (e.g., orientation, locomotion, gesticulation, tactile manipulation, and/or vocalization) that can be judged common or convergent on one or more dimensions (e.g., direction, velocity, tempo, and/or substantive content)." In his book, McPhail further distinguished three phases in the life cycle of temporary gatherings that provide opportunities for collective behaviors to occur: the assembling process, the assembled gathering, and the dispersal process (1991: 153). A gathering in one place at one time is a necessary but insufficient condition. A bunch of people sitting on park benches reading newspapers or fiddling with their smart phones is not collective behavior. The necessary requirement is that their activities are "judged common or convergent," which means that an observer would see two or more people engaging one kind of behavior—such as applauding, laughing, singing, gesticulating, marching—while "one or more additional persons engage in one of more different behaviors" which the observer would judge to be "concerted" (p. 161). Based on extensive fieldwork, McPhail cataloged 40 elementary forms of "collective behavior-in-common" (p. 164). In a large gathering several of those basic behaviors may occur simultaneous at different places within an assembly. Dispersal, the final phase of a gathering, occurs when people leave the common locale for their varied destinations. Most dispersals are routine and unproblematic. In rare emergency dispersals, participants flee from a common site under threats of fire, floods, tornados, explosions, firearm bullets, and other natural or man-made disasters. In between routine and emergency are coerced dispersals "by police, the military, or other agents of social control" (p. 153).

In most of their theoretical and empirical articles, McPhail and Wohlstein used the term collective behavior (e.g., Wohlstein and McPhail 1979; McPhail and Wohlstein 1983, 1986). However, in "Purposive Collective Action," McPhail and Charles Tucker (1990) extensively referred to collective actions in the "nonviolent and violent interactions of demonstrators, counterdemonstrators and social control agents confronting one another in the civil rights and antiwar movements" (p. 82). Their definitions of collective action-in-common and collective action-in-concert are almost identical to the collective behavior definitions used in McPhail's 1991 book (see also McPhail et al. 2006). In practice, the two concepts have essentially become indistinguishable and interchangeable.

Other sociological theorists accepted the conceptual equivalence of the two terms. James Coleman defined the result of a collective action as "an outcome of an event, partially controlled by each of a set of individual actors, will be described as an action or inaction of a collectivity of which these individual actors are members" (1973: 65). The connections between individual actions and collective actions were the core of his decades-long project to reconstruct sociological theory by integrating micro- and macro-level structures and processes (Coleman 1966, 1986). In his magnum opus, *Foundations of Social Theory* (1990), Coleman emphasized collective behaviors that "involve a number of people carrying out same or similar actions at the same time." Those behaviors are "transient or continually changing, not in an equilibrium state." And the actions exhibit some kind of dependency, that is, "individuals are not acting independently" (p. 198). His examples of collective behaviors are quite diverse, ranging from bank runs, panics, riots, the spread of fashions and crazes, to religious frenzies and mass conversions.

Susan Olzak (1989) reviewed methods for using information about collective events to analyze social movements. Her "minimal definition of collective action is that it (a) involves more than one person, and (b) makes claims of agency (or corporate) status" (p. 124). She cited Charles Tilly's definition of claims of agency status as acts that "make a visible claim which, if realized, would affect the interests of some specific person(s) or group(s) outside their own number" (Tilly 1978: 275). His examples of historical agency claims included petitioning, memorializing, and supporting or opposing an enemy of the government. Olzak advocated using event-history analysis methods that enable researchers to measure and compare across diverse forms of collective action. "Thus, information about revolutions consisting of thousands of acts of violence and about one-time confrontations is enumerated and described in terms of number of participants, duration of unrest, magnitude of violence, and other characteristics" (Olzak 1989: 120). Event-history analysis is appropriate for estimating rates of recurrent collective actions, such as riots, demonstrations, revolutions, and social movements, and is "particularly useful for analyzing cycles or waves of protest and violence" (p. 136). The method requires creation of large datasets with information on the timing and sequencing of numerous events.

Pamela Oliver's "core definitional content" of collective action is "a common or shared interest among a group of people" (1993: 272). She reviewed four types of formal theories and models of collective action:

> single-actor models which treat the "group" behavior as given; models of the interdependent aggregation of individual choices into collective action; models of the collective decisions of individuals with different interests; and models of the dynamic interactions among collective actors and their opponents. (p. 272)

At the time Oliver wrote, models aggregating individual choices were showing the greatest growth. She urged greater use of computer simulations in theorizing collective action for two purposes. "The first is to investigate the behavior of a determinate mathematical model under varying initial conditions" (p. 295). And the second is

Monte Carlo simulation of models with probabilistic or stochastic elements. Arguably the proliferation of agent-based models in the early twenty-first century comprises a third purpose.

In recent years, collective action has eclipsed collective behavior as the more frequently used term in such disciplines as anthropology (DeMarrais and Earle 2017; Smith 2017), organizations (Lee et al. 2018; Bridoux and Stoelhorst 2022), political science (Ferguson 2020), sociology (Schweingruber 2021; Putini 2023), social movements (Diani 2011; Ancelovici 2021), and social network analysis (Kapucu and Beaudet 2020; Knoke et al. 2021). Joining that trend, I refer to collective action throughout this book when describing and analyzing the concerted behaviors of people or organizations.

3 A Collective Action Typology

Before attempting to explain collective actions, I identify the diverse phenomena to be investigated. Table 1 classifies three ideal-types that vary in their degree of institutionalization. The attributes of each type may not characterize every real-world instance. Unplanned collective actions are spontaneous events whose time and place are not scheduled in advance by organizers or coordinators. Many are initiated by real or perceived threats to safety and security, for example, natural disasters such as wildfires, hurricanes, or floods. Others are manmade disasters, such as a terrorist attack on a market place, an active shooter inside a school, a crowd surging toward the stage at a rock concert, or partiers fleeing from a nightclub fire.

Organized collective actions involve preplanning then publicizing the place and time when supporters will assemble. Examples include social movements, religious revivals, and election campaigns. Typically, organizational leaders decide on the purpose of a gathering and make the necessary arrangements, such as acquiring permits, notifying the police, and alerting newspaper and television reporters.

Table 1 Attributes of three ideal-types of collective actions

Unplanned collective actions	Organized collective actions	Collective decision-making
• Spontaneous assembly • No leaders or event coordinators • Provoking or triggering incidents • Nonextant network ties • Imitation of behaviors observed in a vicinity • Threshold to join perceived as low-risk • No/low presence of suppressive authorities	• Leaders or coordinators • Formal and informal organizations • Scheduled events to seek a redress of grievances • Participants recruited via social media and personal networks • Event permits issued by public authorities • Police presence to maintain order	• Government institutions • Officials have authority to make binding decisions • Voting rules to decide among alternatives • Creation of public goods & private benefits • Enforcement mechanisms • Limited jurisdictions, balance-of-power

Activists and passive participants are recruited through social and mass media, and via word-of-mouth through personal networks. A single event, such as a rally at city hall plaza, might consist of speeches airing grievances, submission of a petition to the mayor's office, and peacefully dispersing. More complex collective action sequences might link a rally, a march, and a sit-in at three separate locations. Organized protests and demonstrations may devolve into unplanned riots if a few participants begin throwing stones, looting stores, and burning cars. The disturbance spreads as onlookers imitate those actions. Police may intervene to restore order, deploying tear gas and truncheons, potentially arresting scores of rioters. Violence dissolves an organized collective action into a spiral of unplanned property destruction, injuries, and deaths.

Collective decisions are the most institutionalized form of collective action. Coleman identified them as situations "in which a collectivity must make a choice between two (or more) alternatives, though different members may have different preferences" (1973: 31). He argued that the theoretical problem is how to explain such choices in terms of the rational behavior of individuals: "a collective decision ordinarily is disadvantageous to some of those who participate in it, yet they consent to it. Under such conditions, how can a collectivity continue intact, how do collective decisions get made, and when will there be a refusal to consent?" (1966: 625). Democratic institutions, such as constitutions or charters, stipulate the procedures for citizens to elect government officials to assemblies, councils, parliaments, legislatures, and other lawmaking bodies. Some governments permit collective decisions to be made by popular votes on referendums or initiatives. The outcomes of elections are usually accepted by the losers, with Donald Trump's attempt to overthrow the 2020 US presidential results being a notorious exception. Legislative bodies have rules for making collective decisions, with a simple majority vote being the most widely used, while special circumstances may require a super-majority. After a law is enacted, it becomes binding within a jurisdiction although subject to legal challenges and judicial review.

This book is concerned with unplanned and organized collective actions, leaving the study of collective decision-making for another day. Now that the definition and attributes of those collective actions have been delineated, the next section turns to some prominent basic theories of the phenomena.

4 Collective Action Theories

This section reviews three theories or models that explain various types of collective action: flocking, thresholds, and diffusion. Elements of these theories greatly influenced the construction of the agent-based models discussed in this book.

Flocking A murmuration of starlings is hundreds or thousands of birds moving as one across the sky, swirling, swooping, diving, splitting apart and recombining in a dazzling display. Despite their close proximity, starlings do not collide, neither are they following a central leader, nor controlled by global processes. Scientific explanations of such self-organizing systems assume that each bird is autonomous and in direct interaction with only a few others. "Once, the murmuration was consulted as a carrier of the messengers of gods, and later it was the psychic transference of bodies. Now, in the contemporary world, the starling murmuration is understood as individuals following the movements of their nearest neighbors" (Vallee 2021: 83–84). Craig Reynold's (1987) computer simulation of a flock of virtual agents ("boids") mimicked the dynamics of a starling murmuration. It generated coordinated macro-level actions of the flock from three simple behavioral rules at the micro-level of individual bird perceptions: (1) Separation, keeping distance from nearby local flock-mates and obstacles; (2) Alignment, moving towards the average heading and at a velocity matching flock-mates; (3) Cohesion, steering towards the average position (center of mass) of flock-mates. Similar patterns of collective motion can be observed in schools of fish, herds of bison and wildebeests, and swarms of locusts.

Human flocking behavior occurs among pedestrians where walkers use visual cues to adjust their speed of locomotion and angular direction to avoid collisions on crowded sidewalks, train station platforms, or shopping malls and plazas. Experiments on real human subjects embedded in a virtual world of avatars can be explained by "basic properties of interaction as natural consequences of the laws of optics" (Dachner et al. 2022: 8). Optical variables control an individual's heading and speed, with collision avoidance resulting from cancelling optical expansion, and group cohesion maintained by cancelling optical contraction. Collective motion "emerges from cancelling the combined expansion/contraction and the angular velocity of neighbours" (p. 8). The number of nearest neighbors influencing a walker's motion is not determined by a fixed distance but by crowd density. In dense crowds, complete visual occlusion may occur within a range of five meters (16 feet). "In summary, we conclude that the local interactions underlying collective motion have a lawful basis in the visual coupling between neighbours." Other investigations of human flocking behavior deployed computer simulations to analyze self-organization from local interactions among persons without central algorithms or designated leaders (Belz et al. 2013; Frey and Goldstone 2018; Warren et al. 2024).

Thresholds Mark Granovetter's (1978) threshold model of collective behavior assumed an assembly of rational individuals—able to calculate the costs and benefits of an action—each observing what everyone else at the gathering is doing before deciding whether to join in or refrain from the action. "A person's threshold for joining a riot is defined here as the proportion of the group he would have to see join before he would do so" (1978: 1422). The distribution of thresholds within the group determines whether a riot erupts or peters out. If a crowd has many low-threshold members, a few instigators could trigger a cascade as the swelling number of participants sequentially persuades higher-threshold members to join. But if an assembly is comprised of people with predominantly high thresholds, a handful of initiators would not suffice to induce a wave of imitators to sweep across the crowd.

An important factor affecting the success or failure of collective action is the social structure of the gathering. "By 'social structure' I mean here only that the influence any given person has on one's behavior may depend upon the relationship" (p. 1429). For example, if the influence of friends is twice as high as that of strangers, activation is much more likely when a bystander sees one or more friends joining a riot than when a similar number of strangers join. Another important factor is the spatial/temporal dispersion of the action:

> ... in cities where a fairly constant distribution of riot thresholds exists the outcome of one crowd may nevertheless differ radically from that generated by another, at some earlier or later time, for reasons that have nothing to do with differences between the occasions or intervening events but involve only sampling variability (p. 1431).

In addition to rioting, Granovetter contended that threshold models could be fruitfully applied to such diverse collective behaviors as going to college, strikes, migration, markets, voting, leaving social occasions, conformity, innovation adoption, diffusion of rumors and diseases, and residential segregation.

The dynamics of residential segregation exemplify rapid change when thresholds exceed a tipping point. Thomas Schelling (1969, 1971, 1978) demonstrated that "slight-but-not-malicious" in-group preferences (homophilia)—people prefer to move into residential neighborhoods that are occupied by people like themselves—produce the unintended consequence that communities become highly segregated, for example, by race, ethnicity, income, or politics. The concept of rapidly transforming collective actions, such as US "white flight" from inner cities to suburbs, were widely popularized by journalist Malcolm Gladwell's best-seller, *The Tipping Point* (2000), "the magic moment when an idea, trend, or social behavior crosses a threshold, tips, and spreads like wildfire" (publisher's description; see also Gladwell 2024). Recent refinements in threshold models of collective action include Wiedermann et al. (2020), Kaaronen and Strelkovskii (2020), Macy and Evtushenko (2020), Macy et al. (2021), Milkoreit (2023), and Warren et al. (2024).

Diffusion An innovation is a change in technology, practice, belief, or routine. Most innovations are minor competence-enhancing changes that easily fit into existing thoughts and behaviors. Much rarer competence-destroying breakthroughs can threaten the status quo by unleashing "gales of creative destruction" (Schumpeter 1926) that radically transform industries, markets, or political systems. Diffusion refers to the spreading of information, items, or activities across a social system from initiating sources to receiving adopters via communication, influence, threat, or persuasion.

> We use the term "practice" to denote the diffusing item, which might be a behavior, strategy, belief, technology, or structure. Diffusion is the most general and abstract term we have for this sort of process, embracing contagion, mimicry, social learning, organized dissemination, and other family members (Strang and Soule 1998: 266).

A classic study of the adoption of a new drug (tetracycline) by 125 physicians in four Illinois towns found that early prescription writing depended more on network relations than mass media advertising:

> … networks of doctor-to-doctor contacts operated most powerfully in the first 5 months after the release of the new drug. … The discussion network and the advisor network showed most pair-simultaneity at the very beginning and then progressively declined. The friendship network … appears to reach maximum effectiveness later (Coleman et al. 1957).

Reanalyzing the Illinois physician data, Thomas Valente applied a network threshold model that assumed "behavioral contagion through direct network ties" (1996: 85). Granovetter's collective behavior threshold model assumed thresholds are the proportion of adopters in a social system prior to an individual's adoption. In contrast, Valente's model did not require individuals to monitor everyone in a system. Instead, thresholds are measured at the egocentric level, for example, the number of physicians (alters) in an ego doctor's friendship, advice, and discussion network. Exposure to an innovation at a given time is calculated by dividing the number of alters who have adopted by the total size of ego's personal network. The threshold is the exposure at the time-of-adoption. The proportion of adopters generally increases over time as more and more alters adopt and "varies across individuals according to the adoption behavior of their network partners" (p. 73). The early adopters have lower thresholds than the laggards. Recent research on innovations diffusing via networks includes Zhang and Centola (2019), Valente and Yon (2020), Aidt et al. (2022), Leite (2022), Mueller and Ramkumar (2023), and Hu et al. (2024).

5 Overview of Chapters

Chapter "Modeling" describes the methodology of agent-based models, emphasizing the NetLogo programming language used in this book. It discusses the advantages and limitations of ABMs for running experiments to test hypotheses derived from collective action theories. The heart of this book are six chapters that: review social science theories and research on a specific type of collective action; summarize a few prior ABM models on the topic; present a simple ABM illustrating a core feature of the phenomenon; and discuss some directions for further research. Chapter "Infecting" is about the spread of contagious diseases, exemplified by the differential consequences of compliance or avoidance of social distancing and face masking in the absence of a vaccine. What was the relation between US political partisanship and Covid-19 deaths? Chapter "Crushing" examines how overcrowding in a confined space—rushing toward a religious shrine or fleeing from a nightclub fire—can lead to crowd collapse and death by crushing. Are people with network ties to others in a crowd more vulnerable than strangers?

Chapter "Rioting" depicts rioting in unplanned assemblies where people with low risk-thresholds join the violent acts of instigators. Can the presence of civilian peacemakers quell the spread of disorderly conduct? Chapter "Recruiting" investigates how personal networks affect mobilization of participants for social movement events such as marches, rallies, and sit-ins. How does the social structure of small networks affect recruitment success or failure? Chapter "Insurrecting" models an insurrection, specifically the January 6, 2021, attack on the US Capitol by organized groups heeding President Donald Trump's call to prevent Congressional certification of the Electoral College votes. Would a larger number of police or national guard have resulted in a different outcome? Chapter "Insurging" looks at insurgencies where an armed guerrilla force fights soldiers of a central government for control of contested terrain. Is the outcome more affected by the relative sizes of the contending parties or by the lethality of their weapons? Finally, chapter "Anticipating" attempts to peer into the future of agent-based models of collective action, with speculations about the impact of artificial intelligence.

References

Aidt T, Leon-Ablan G, Satchell M (2022) The social dynamics of collective action: evidence from the diffusion of the swing riots, 1830–1831. J Polit 84:209–225
Ancelovici M (2021) Conceptualizing the context of collective action: an introduction. Soc Mov Stud 20:125–138
Belz M, Pyritz LW, Boos M (2013) Spontaneous flocking in human groups. Behav Process 92:6–14
Bridoux F, Stoelhorst JW (2022) Stakeholder governance: solving the collective action problems in joint value creation. Acad Manag Rev 47:214–236
Caputo A, Craft W, Gilbert C (2020) 'The precinct is on fire': what happened at Minneapolis' 3rd precinct—and what it means. APMreports, 30 June
Coleman JS (1966) Foundations for a theory of collective decisions. Am J Sociol 91:615–627
Coleman JS (1973) The mathematics of collective action. Aldine, Chicago
Coleman JS (1986) Social theory, social research, and a theory of action. Am J Sociol 91:1309–1335
Coleman JS (1990) Foundations of social theory. Harvard University Press, Cambridge, MA
Coleman JS, Katz E, Menzel H (1957) The diffusion of an innovation among physicians. Sociometry 20:253–270
Dachner GC, Wirth TD, Richmond E, Warren WH (2022) The visual coupling between neighbours explains local interactions underlying human 'flocking'. Proc R Soc B 289(1970):20212089
DeMarrais E, Earle T (2017) Collective action theory and the dynamics of complex societies. Annu Rev Anthropol 46:183–201
Diani M (2011) Social movements and collective action. In: Scott J, Carrington P (eds) The SAGE handbook of social network analysis. Sage Publications, Thousand Oaks, pp 223–235
Ferguson WD (2020) The political economy of collective action, inequality, and development. Stanford University Press, Stanford
Frey S, Goldstone RL (2018) Cognitive mechanisms for human flocking dynamics. J Comput Soc Sci 1:349–375
Gladwell M (2000) The tipping point: how little things can make a bid difference. Little, Brown, Boston
Gladwell M (2024) Revenge of the tipping point: overstories, superspreaders, and the rise of social engineering. Little, Brown, Boston.
Granovetter M (1978) Threshold models of collective behavior. Am J Sociol 83:1420–1443. Little, Brown, Boston
Hu F, Qiu L, Wei S, Zhou H, Bathuure IA, Hao H (2024) The spatiotemporal evolution of global innovation networks and the changing position of China: a social network analysis based on cooperative patents. R&D Manag 54:574–589
Kaaronen RO, Strelkovskii N (2020) Cultural evolution of sustainable behaviors: pro-environmental tipping points in an agent-based model. One Earth 2:85–97
Kapucu N, Beaudet S (2020) Network governance for collective action in implementing United Nations sustainable development goals. Admin Sci 10(4):100

Knoke D, Diani M, Hollway J, Christopoulos D (2021) Multimodal political networks. Cambridge University Press, Cambridge
LeBon G (1895) Psychologie des Foules. Alcan, Paris
Lee BH, Struben J, Bingham CB (2018) Collective action and market formation: an integrative framework. Strateg Manag J 39:242–266
Leite E (2022) Innovation networks for social impact: an empirical study on multi-actor collaboration in projects for smart cities. J Bus Res 139:325–337
Macy MW, Evtushenko A (2020) Threshold models of collective behavior II: the predictability paradox and spontaneous instigation. Sociol Sci 7:628–648
Macy MW, Ma M, Tabin DR, Gao J, Szymanski BK (2021) Polarization and tipping points. Proc Natl Acad Sci 118(50):e2102144118
McPhail C (1991) The myth of the madding crowd. Aldine de Gruyter, New York
McPhail C, Tucker CW (1990) Purposive collective action. Am Behav Sci 34:81–94
McPhail C, Wohlstein RT (1983) Individual and collective behaviors within gatherings, demonstrations, and riots. Annu Rev Sociol 9:579–600
McPhail C, Wohlstein RT (1986) Collective locomotion as collective behavior. Am Sociol Rev 51:447–463
McPhail C, Schweingrube DS, Ceobanu A (2006) Purposive collective action. In: McClelland KA, Fararo TJ (eds) Purpose, meaning, and action: control systems theories in sociology. Palgrave Macmillan, New York, pp 57–83
Milkoreit M (2023) Social tipping points everywhere? Patterns and risks of overuse. Wiley Interdiscip Rev Clim Chang 14(2):e813
Mueller M, Ramkumar S (2023) Signed networks-the role of negative links for the diffusion of innovation. Technol Forecast Soc Chang 192:122575
Oliver PE (1993) Formal models of collective action. Annu Rev Sociol 19:271–300
Olzak S (1989) Analysis of events in the study of collective action. Annu Rev Sociol 15:119–141
Penrod J, Sinner CJ (2020) Buildings damaged in Minneapolis and St. Paul after Rios. Minneapolis Star Tribune, 13 July
Putini A (2023) Collective action, urban spaces, and common goods: the concept of informality from a sociological perspective. In: Ferroni MV, Galdini R, Ruocco G (eds) Urban informality: a multidisciplinary perspective. Springer International Publishing, New York, pp 21–39
Reynolds CW (1987) Flocks, herds, and schools: a distributed behavioral model. Comput Graph 21(4):25–34
Schelling TC (1969) Models of segregation. Am Econ Rev 59:488–493
Schelling TC (1971) Dynamic models of segregation. J Math Sociol 1:143–186
Schelling T (1978) Micromotives and macrobehavior. Norton, New York
Schumpeter JA (1926) Theorie der wirtschaftlichen Entwicklung, 2nd edn. Duncker & Humbolt, München/Leipzig
Schweingruber D (2021) The capitol breach: perspective from the sociology of collective action. Dyn Asymmetric Confl 14:110–120
Smith MG (2017) Corporations and society: the social anthropology of collective action. Routledge, London
Strang D, Soule SA (1998) Diffusion in organizations and social movements: from hybrid corn to poison pills. Annu Rev Sociol 24:265–290
Tilly C (1978) From mobilization to revolution. Addison Wesley, Reading
Valente TW (1996) Social network thresholds in the diffusion of innovations. Soc Networks 18:69–89
Valente TW, Vega Yon GG (2020) Diffusion/contagion processes on social networks. Health Educ Behav 47:235–248
Vallee M (2021) Animal, body, data: starling murmurations and the dynamic of becoming in-formation. Body Soc 27(2):83–106
Warren WH, Benjamin Falandays J, Yoshida K, Wirth TD, Free BA (2024) Human crowds as social networks: collective dynamics of consensus and polarization. Perspect Psychol Sci 19:522–537

Wiedermann M, Keith Smith E, Heitzig J, Donges JF (2020) A network-based microfoundation of Granovetter's threshold model for social tipping. Sci Rep 10:11202

Wohlstein RT, McPhail C (1979) Judging the presence and extent of collective behavior from film records. Soc Psychol Q 42:76–81

Zhang J, Centola D (2019) Social networks and health: new developments in diffusion, online and offline. Annu Rev Sociol 45:91–109

Modeling

Keywords Agent-based model · Agents · Cellular automata · Emergence · Environments · Game of Life · Moore neighborhood · NetLogo · Rules · Schedulers · Sensitivity analysis · von Neumann neighborhood

1 Introduction

This chapter describes the methods used in this book to simulate collective actions. It begins with the roots in cellular automata, a grid of cells that change states over time through the influence of neighboring cells. The next section discusses social network analysis. It then turns to agent-based models (ABMs), a computational method for simulating the interactions among autonomous agents through space and time to change outcomes at the systemic level. The chapter concludes with an overview of NetLogo, an accessible programming language used in this book's six substantive chapters to create ABMs of collective action. Social scientists interested in learning more about ABM applications in their disciplines could consult overviews of such disciplines as criminology (Groff et al. 2019), economics (Axtell and Doyne Farmer 2022), diffusion of innovation (Zhang and Vorobeychik 2019), social networks (Namatame and Chen 2016), management (Wall 2015), migration (Klabunde and Willekens 2016), organizations (Secchi 2015), political science (De Marchi and Page 2014; Dacrema and Benati 2020), public health (Tracy et al. 2018), social psychology (Jackson et al. 2017), sociology (Macy and Willer 2002; Edelmann et al. 2020), and supply chains (Khajouei et al. 2021). Although the substantive issues vary substantially, all share the principle that outcomes at the systemic level of analysis emerge from the actions of autonomous agents at the local level of analysis.

D. Knoke, *Network Collective Action*, Lecture Notes in Social Networks,
https://doi.org/10.1007/978-3-031-86199-4_2

2 Cellular Automata Models

Architects, engineers, fire departments, and public safety agencies all have keen practical interests in understanding evacuations from burning buildings. The earliest research models of crowd behavior applied principles drawn from particle physics and fluid dynamics (gases and liquids) to construct differential equations treating crowds as flows of points through fields in continuous time (Li et al. 2019). In the final decades of the twentieth century, cellular automata computational models of complex, self-organizing systems spread rapidly in physics, biology, chemistry, and related technological fields, such as materials sciences and robotics (Clarke 2021). Social science applications included land-use planning, urban growth, and traffic flows (Hegselmann and Flache 1996). The origin of these models lies in cellular automata concepts introduced after the end of the Second World War.

Developed through discussions between polymath John von Neumann and mathematician Stanislaw Ulam in the 1940s, a cellular automaton is a cell in a discrete two-dimensional lattice (similar to a checkerboard) whose behavior depends on the actions or states of its neighbors (von Neumann 1951, 1966). Automata analysts typical posit one of two types of adjacent neighborhoods. As shown in Fig. 1, a von

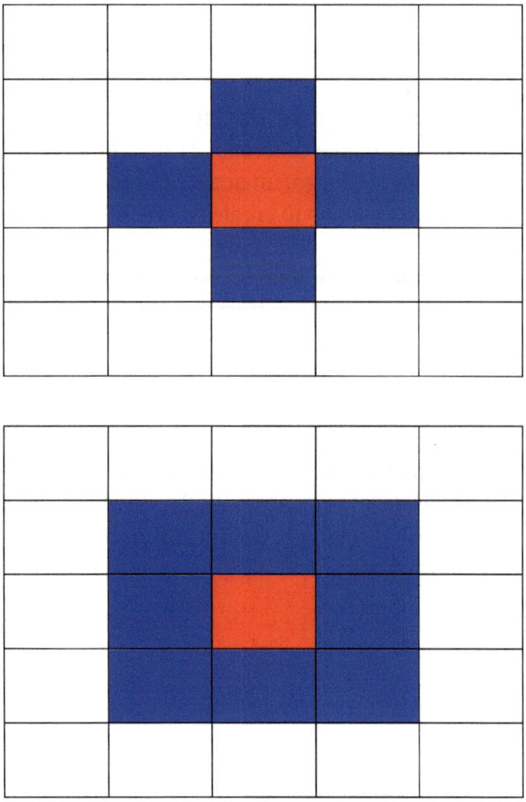

Fig. 1 Von Neumann and Moore neighborhoods for cellular automata

Neumann neighborhood is comprised of the four blue cells with an edge abutting the red reference automaton (orthogonally adjacent, corresponding to north, east, south, and west on a map). In turn, each of the four blue cells has its own von Neumann neighborhood, one cell of which is the red automaton. A Moore neighborhood, named after mathematician Edward F. Moore, adds the four cells whose corners meet the reference cell's four corners (diagonally adjacent). In both examples, the neighborhood has radius $r = 1$ because only one layer of cells surrounding the reference cell is taken into account. Moore neighborhoods could be expanded using a larger radius that encompasses additional layers around the reference cell. Every cell in a neighborhood has an initial set of status information, typically a binary attribute such as color (e.g., white, black), number (0, 1), or well-being (dead, alive).

Action begins when all automata interact with their neighbors and simultaneously change status in discrete time based on a simple set of rules. Perhaps the most popular illustration is mathematician John Conway's Game of Life, using a Moore neighborhood, where cells live or die depending on local rules about birth, death, and survival:

1. A live cell with fewer than two live neighbors dies (underpopulation)
2. A live cell with two or three live neighbors remain alive
3. A live cell with more than three live neighbors dies (overpopulation)
4. A dead cell with exactly three live neighbors comes alive (reproduction)

If none of the conditions are met, the automaton remains in its current state until further time-steps alter its neighborhood. The rules apply to all cells at every time-step during the Game. Depending on the initial configuration, a stable pattern may emerge that oscillates in an endlessly repeating cycle. Cellular automata demonstrate how applying a few simple rules of local interactions may generate complex macro-level phenomena. Martin Gardner popularized Conway's Game for a lay audience in his *Scientific American* column: "Because of its analogies with the rise, fall and alternations of a society of living organisms, it belongs to a growing class of what are called 'simulation games' – games that resemble real-life processes" (Gardner 1970: 120).

Cellular automata models of social behavior began appearing in the late twentieth century (e.g., Itami 1994; Leydesdorff 1995; Hegselmann and Flache 1996). Many depicted crowd behavior under duress, for example, people fleeing a fire through exits in a single room, while others investigated evacuations from more complex structures, such as high-rise office and apartment buildings, ships, and trains, or from facilities with large mobility-impaired populations, such as nursing homes and hospitals (e.g., Pelechano and Malkawi 2008; Chen and Han 2015; Wang et al. 2021). One CA model simulated a fire spreading to neighboring cells from a random origin while randomly distributed pedestrians all share the goal of finding an escape route to the nearest of four exits (Alidmat et al. 2015). Their movements were "represented as 'chaotic', mimicking panic egress behaviors during a fire evacuation. This model includes a fire circular front shape based on the spiral movement technique" (p. 293). Another CA model used the grid attraction index values of an occupied cell's eight Moore neighbors to select direction of movement away from the fire and toward the nearest exit (Yang et al. 2014). Occupants had unequal access to target cells, with younger men moving onto unoccupied cells before older women. In these and other CA models, realism is hindered because occupied cells represent both locations in physical space and actor attributes. Another limitation is that no cell can be occupied by more one person at a time, which means that overcrowding and crushing outcomes are more difficult to depict. Fortunately, actor-based computational models of social behavior separate these two features and allow for more complex, nuanced simulations.

3 Agent-Based Models

While acknowledging that a universally accepted definition of agent-based modeling and simulation (ABMS) is unlikely, Charles Macal (2016: 149–151) instead proposed four increasingly elaborate definitions:

Definition 1:
An *individual ABMS* is one in which the agents in the model are represented individually and have diverse characteristics.
Definition 2:
An *autonomous ABMS* is one in which the individual agents have internal behaviors that allow them to be autonomous, able to sense whatever condition occurs within the model at any time, and to act on the appropriate behavior in response.
Definition 3:
An *interactive ABMS* is one in which autonomous agents interact with other agents and with the environment.
Definition 4:
An *adaptive ABMS* is one in which the interacting, autonomous agents change their behaviors during the simulation, as agents learn, encounter novel situations, or as populations adjust their composition to include larger proportions of agents who have successfully adapted.

Macal argued that in creating or analyzing an agent-based model, researchers should ask how its agents are characterized in terms of the four key concepts of individuality, autonomy, interactivity, and adaptability. Most of the ABMs discussed in this book can be considered interactive models in which autonomous agents interact with one another and the environment. The concluding chapter briefly speculates the potential for artificial intelligence to be integrated with ABMs, producing adaptive agent-based models in which agents learn from their experiences and adjust their actions.

Four basic conceptual building blocks of ABMs are agents, rules, environments (Choi and Park 2021), and schedulers (Masad and Kazil 2015). **Agents** are often conceptualized as representing people, although they can be other types of entities such as groups, organizations, and nations. Agents have two elements: attributes and actions. Attributes may differ among individual agents or subsets. Attributes may be fixed demographic characteristics, such as gender, race, and age, or they could change during a simulation, such as money, information, and network contacts. Agent actions are behaviors that agents may initiate or suffer from, such as moving across an environment, interact with other agents, kill or be killed. As with attributes, potential actions may vary among individuals or subsets of agents. For example, the speed of movement might be faster for younger agents than older ones. Or agents are more likely to communicate and collaborate with others sharing the same or similar attributes (homophily preference).

Rules govern the actions that agents take during specific encounters (Macy and Willer 2002). ABMs explore the macro-level consequences of agents following simple decision-making rules at the micro-level, sometimes producing surprising or unexpected systemic consequences. For example, traffic jams occur when agents driving cars on a highway follow two rules: (1) If there's a car close ahead, slow down; (2) If no cars are ahead, speed up (Wilensky 1997). Although all cars move in one direction on a circular loop, resembling an auto racing track, when some cars are randomly clustered together, they slow down, causing others behind them to slow down and a traffic jam begins. Although the cars travel forward on the track, the jam moves backward like a wave: the collective behavior is different than the actions of individual drivers. Joshua Epstein's (2002) ABM of aggrieved agents rebelling against an oppressive government used such rules as "If grievance minus net risk exceeds a nonnegative threshold, be active; otherwise, be quiet." His model exhibited an unexpected emergence of individually deceptive behavior: when large numbers of police come into a neighborhood, privately aggrieved agents hide their feelings, but whenever the cops are away, the aggrieved citizens burst into openly rebellious actions. ABM designers strive to specify sets of rules according to the theoretical or empirical actions they wish to simulate.

Environments are typically two-dimensional spatial lattices, resembling cellular automata, on which agent actions unfold. Environments range from featureless blank fields to elaborate natural landscapes with rivers, forests, and mountains to detailed conurbations of buildings, factories, docks, airports, stadiums, roads, and bridges. Those elements may act as barriers that constrain agents' spatial movements and may require rules that specify feasible agent-environment interactions. Because more than one agent can simultaneously occupy the same environmental location, a safety rule might stipulate that agents move away from perceived dense locations to avoid crushing injuries or death. Some ABMs created during the Covid-19 global pandemic examined the spread of infections across populations under imposed and relaxed masking, social distancing, and lockdown mandates in dense urban areas and sparse rural regions (e.g., Giacopelli 2021; Grinberger and Felsenstein 2023).

Schedulers are computer routines that control agent activation sequences and keep track of time during a simulation (Masad and Kazil 2015). Models often distinguish between a step ("tick") and the activation of a single agent. In *synchronous activation*, all agents act simultaneously. In a *uniform activation* all agents are activated in the same order at every step. *Random schedulers* also activate all agents at every step, but in random sequences. Some schedulers keep track of a simulation's clock by counting how many steps (ticks) have elapsed, while others track simulated clock time, such as minutes, hours, or days. Because ABMs can run indefinitely, researchers should always specify an end condition for terminating a run; for example, whenever an equilibrium is reached where no further micro-level changes are possible (e.g., all agents have stopped moving or have expired).

A fundamental purpose of ABMs is to investigate **emergence**: how global properties of macro-systems "emerge from the bottom up, through local interactions" (Macy and Willer 2002: 147). In classical philosophical British emergentism, the relationship was absolutely inexplicable: the whole cannot be logically deduced from its parts (Epstein 1999: 53). In contrast, "to the agent-based modeler, it is precisely the generative sufficiency of the parts (the microspecification) that constitutes the whole's explanation! In this particular sense, agent-based modeling is reductionist" (p. 55). The method offers social scientists ways to explain how individual-level rules produce systemic-level regularities. How stable are the macroscopic configurations arising from the local interactions among autonomous, rule-following agents? Do small fluctuations in the latter generate large, unexpected alterations of the former?

Agent-based modeling is a valuable tool for operationalizing and testing multilevel social theories. The imprecise assumptions, concepts, and propositions of a verbal theory must be translated into explicit ABM mechanisms linking agent micro-actions to emergent macro-consequences. Model outcomes then feed back into clarification and refinement of the theory. Working in tandem, theory and model ideally improve one another. "ABMs can function as prostheses for the imagination, increasing our power to develop theory" (Smaldino et al. 2015: 311). Data produced by an ABM may be incongruent with real-world data from other sources, such as sample surveys, censuses, experiments, ethnographies, and archived documents. In such instances, model modifications may identify elements needed to bring theory, model, and data into closer agreement.

Another benefit of agent-based modeling is its ability to perform **sensitivity analyses**. "Sensitivity involves the effect on output (generated macrostructure) of small changes in input (microspecification)" (Epstein 1999: 52). Traditional one-factor-at-a-time (OFAT) testing selects a base parameter setting, varies one focal parameter, and keeps all other parameters fixed (Ten Broeke et al. 2016). First, a researcher runs one set of ABM parameters hundreds or thousands of times and tabulates the mean and standard deviations of various outcome measures. Next, after slightly altering the focal parameter's value, run that version of the model hundreds or thousands of times, again tabulating the outcomes. Repeat across a wide range of focal parameter values. Graph the experimental results to assess whether the response is linear or nonlinear and whether any tipping point generates large or unexpected systemic changes. Do the experiments warrant reassessment of the theory, the model, or both?

Critics of agent-based models raise numerous objections, ranging from incorrect parameterization, to inability to validate and verify accuracy, to arbitrariness (Rand and Stummer 2021: 433–435; see also An et al. 2020; Choi and Park 2021). All models are simplifications of reality and many share the limitations of ABMs; for examples, the rational utility-maximizing consumer in economics and the linear regression equations of status attainment path models in sociology. Agent-based modeling is a work in progress, whose current limitations should not be a reason to abandon efforts at improving them. As statistician George Box remarked, "Remember that all models are wrong; the practical question is how wrong do they have to be to not be useful" (Box and Draper 1987: 74). Agent-based modelers should constantly strive to develop more useful models.

3.1 NetLogo

To design and run ABMs, social science researchers must learn a programming language. This section provides a brief overview of NetLogo, the software used in this book. The six featured ABMs in chapters "Infecting", "Crushing", "Rioting", "Recruiting", "Insurrecting" and "Insurging" are hosted on SpringerLink as electronic supplementary material. To use those models, readers should visit NetLogo at <https://ccl.northwestern.edu/netlogo/> and download the NetLogo software. The programs are also available on the NetLogo Modeling Commons <http://modeling-commons.org/account/login>.

Dozens of ABM platforms are available, including AnyLogic (Garifullin et al. 2007), Mason (Luke et al. 2005), Repast HPC (Collier and North 2011); and Swarm (Minar et al. 1996). Assessments of these platforms and many other general and specialized toolkits seek to guide newcomers to select the softwares most suitable to their needs and skills (e.g., Nikolai and Madey 2009; Pal et al. 2020; Antelmi et al. 2022). Numerous publications and online Websites offer tutorials for novices (e.g., Macal and North 2010; Ghaffarzadegan et al. 2024; Sayama 2024).

NetLogo is an open-source programming language and integrated development environment (IDE) for agent-based modeling. It was designed in 1999 by Uri Wilensky, director of Northwestern University's Center for Connected Learning and Computer-Based Modeling (CCL). He intended it to be accessible by high school students and professionals with limited computing skills in programming. NetLogo software can be downloaded for free, along with manuals, dictionaries, tutorials, and an online Modeling Commons, an archive for sharing and discussing more than a thousand user-contributed NetLogo ABMs spanning the natural sciences, art, philosophy, and social sciences (Wilensky and Rand 2015; Hamill and Gilbert 2016; Gilbert 2020). An indispensable document for learning how to create the ABMs presented in this book was the *NetLogo User Manual 6.4.0* (Wilensky 2023). It informed the descriptions below.

A NetLogo file has three tabs: (1) a Code editor for writing a program; (2) an Info editor for descriptions of what the program does; and (3) an Interface with controls for setting model parameters, a window to display the interactions of agents and the environment at each time step (tick), and plotters to graph and monitors to tabulate the results of a program run. A useful auxiliary tool is the BehaviorSpace, which summarizes the results of multiple model runs and exports the data as an

Excel spreadsheet for statistical analysis. A very important contribution to the programs in this book is the NetLogo Nw Extension, invoked by adding the extension keyword at the beginning of the Code tab. This extension enables the programmer to create network links among a set of agents. It also enables the calculation of such network measures as density, path length, and centrality.

Figure 2 illustrates some NetLogo Interface features for an ABM depicting social movement recruitment on a network (chapter "Recruiting" has a more detailed discussion). Clicking the "Setup" button at the top left generated a network with the parameters chosen in the four green sliders directly underneath: 50 agents with an average of 4 connections to other agents, one green instigator, and a resistance threshold of 20 (i.e., probability = .20). In this model run, 12 red resisters are randomly scattered among 37 neutral blue agents. These neutral agents could become movement supporters (turning green) if they receive an "invitation" to join via a direct tie to a green movement member. But when no lines connecting green and blue agents remain, typically because red resisters blocked the path, the run ended. In Fig. 2, that end came at 5 ticks because the red resister at the center of the diagram was a linch pin between the lower and upper portions of the network. The plot at the center left charts the evolution of the network as blue agents were successively persuaded to join the movement. As summarized by the three monitors at the bottom left, by the end of the run 42% of the agents had become green movement

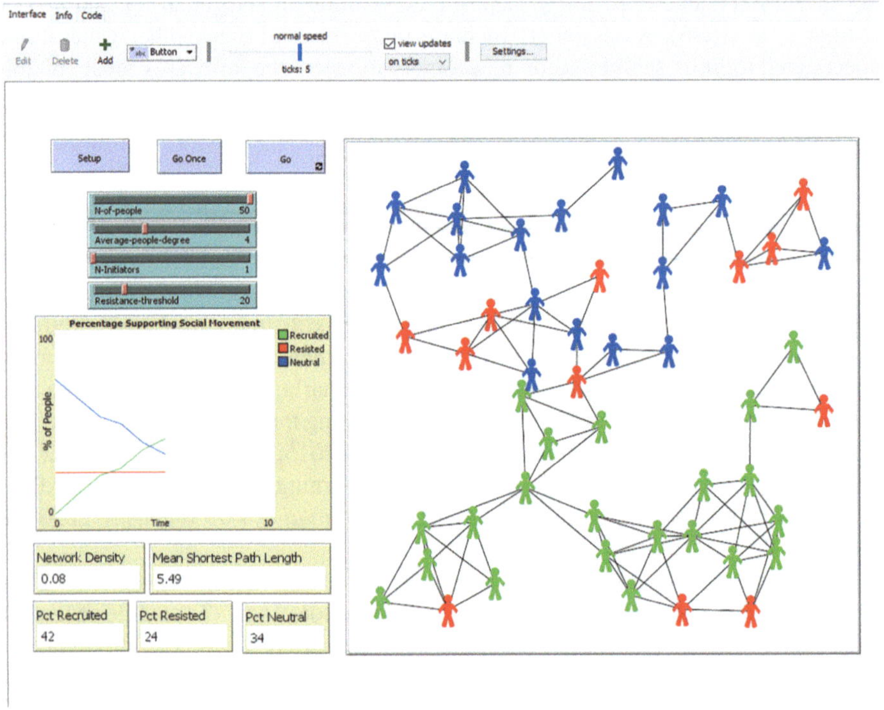

Fig. 2 NetLogo Interface screen capture of social movement recruitment on a network ABM

supporters, 24% were red resisters, and 34% were blue neutrals. Chapter "Recruiting" discusses some experiments with varied social network parameters.

References

Alidmat OKA, Hassan FH, Khader AT (2015) Cellular automata model for pedestrian evacuation in fire spreading conditions. In: Proceedings of the 5th international conference on computing and informatics, pp 293–299

An L, Grimm V, Turner II BL (2020) Meeting grand challenges in agent-based models. J Artif Soc Soc Simul 23

Antelmi A, Cordasco G, D'Ambrosio G, De Vinco D, Spagnuolo C (2022) Experimenting with agent-based model simulation tools. Appl Sci 13:13

Axtell RL, Doyne Farmer J (2022) Agent-based modeling in economics and finance: past, present, and future. J Econ Lit:1–101

Box GEP, Draper NR (1987) Empirical model-building and response surfaces. Wiley, New York

Chen M, Han D (2015) Multi-grid model for crowd's evacuation in ships based on cellular automata. Pol Marit Res 22:75–81

Choi T, Park S (2021) Theory building via agent-based modeling in public administration research: vindications and limitations. Int J Public Sect Manag 34:614–629

Clarke KC (2021) Cellular automata and agent-based models. In: Handbook of regional science. Springer, Berlin/Heidelberg, pp 1751–1766

Collier N, North M (2011) Repast HPC: a platform for large-scale agent-based modeling. Large-Scale Comput:81–109

Dacrema E, Benati S (2020) The mechanics of contentious politics: an agent-based modeling approach. J Math Sociol 44:163–198

De Marchi S, Page SE (2014) Agent-based models. Annu Rev Polit Sci 17:1–20

Edelmann A, Wolff T, Montagne D, Bail CA (2020) Computational social science and sociology. Annu Rev Sociol 46:61–81

Epstein JM (1999) Agent-based computational models and generative social science. Complexity 4(5):41–60

Epstein JM (2002) Modeling civil violence: an agent-based computational approach. Proc Natl Acad Sci 99:7243–7250

Gardner M (1970) The fantastic combinations of John Conway's new solitaire game 'life'. Sci Am 223:120–123

Garifullin M, Borshchev A, Popkov T (2007) Using AnyLogic and agent-based approach to model consumer market. In: Proceedings of the 6th EUROSIM congress on modelling and simulation. Ljubljana, Slovenia, pp 1–5

Ghaffarzadegan N, Majumdar A, Williams R, Hosseinichimeh N (2024) Generative agent-based modeling: an introduction and tutorial. Syst Dyn Rev 40:e1761

Giacopelli G (2021) A full-scale agent-based model to hypothetically explore the impact of lockdown, social distancing, and vaccination during the COVID-19 pandemic in Lombardy, Italy: model development. J Med Internet Res 2(3):e24630

Gilbert N (2020) Agent-based models, 2nd edn. Sage Publications, Thousand Oaks

Grinberger AY, Felsenstein D (2023) Agent-based simulation of COVID-19 containment measures: the case of lockdowns in cities. Lett Spat Resour Sci 16:10

Groff ER, Johnson SD, Thornton A (2019) State of the art in agent-based modeling of urban crime: an overview. J Quant Criminol 35:155–193

Hamill L, Gilbert N (2016) Agent-based modelling in economics. Wiley, Chichester

Hegselmann R, Flache A (1996) Understanding complex social dynamics: a plea for cellular automata based modelling. J Artif Soc Soc Simul 1(3)

Itami RM (1994) Simulating spatial dynamics: cellular automata theory. Landsc Urban Plan 30(1–2):27–47
Jackson JC, Rand D, Lewis K, Norton MI, Gray K (2017) Agent-based modeling: a guide for social psychologists. Soc Psychol Personal Sci 8:387–395
Khajouei MH, Hosseini S, Pilevari N, Radfar R, Mohtashami A (2021) Complex adaptive systems, agent-based modeling and supply chain network management: a systematic literature review. J Ind Eng Manag Stud 8:54–92
Klabunde A, Willekens F (2016) Decision-making in agent-based models of migration: state of the art and challenges. Eur J Popul 32:73–97
Leydesdorff L (1995) The operation of the social system in a model based on cellular automata. Soc Sci Inf 34:413–441
Li Y, Chen M, Dou Z, Zheng X, Cheng Y, Mebarki A (2019) A review of cellular automata models for crowd evacuation. Physica A Stat Mech Appl 526:120752
Luke S, Cioffi-Revilla C, Panait L, Sullivan K, Balan G (2005) Mason: a multiagent simulation environment. SIMULATION 81:517–527
Macal CM (2016) Everything you need to know about agent-based modelling and simulation. J Simul 10:144–156
Macal CM, North MJ (2010) Tutorial on agent-based modeling and simulation. J Simul 4:151–162
Macy MW, Willer R (2002) From factors to actors: computational sociology and agent-based models. Annu Rev Sociol 28:143–166
Masad D, Kazil JL (2015) Mesa: an agent-based modeling framework. In: Proceedings of the 14th Python in science conference, pp 51–58
Minar N, Burkhart R, Langton C, Askenazi M (1996) The swarm simulation system: a toolkit for building multi-agent simulations. Sante Fe Institute Working Paper 1996-06-042
Namatame A, Chen S-H (2016) Agent-based modeling and network dynamics. Oxford University Press, Oxford
Nikolai C, Madey G (2009) Tools of the trade: a survey of various agent based modeling platforms. J Artif Soc Soc Simul 12(2)
Pal C-V, Leon F, Paprzycki M, Ganzha M (2020) A review of platforms for the development of agent systems. arXiv preprint:arXiv:2007.08961
Pelechano N, Malkawi A (2008) Evacuation simulation models: challenges in modeling high rise building evacuation with cellular automata approaches. Autom Constr 17:377–385
Rand W, Stummer C (2021) Agent-based modeling of new product market diffusion: an overview of strengths and criticisms. Ann Oper Res 305:425–447
Sayama H (2024) "19: agent-based models." https://math.libretexts.org/Bookshelves/Scientific_Computing_Simulations_and_Modeling/Introduction_to_the_Modeling_and_Analysis_of_Complex_Systems_. Accessed 6 June 2024
Secchi D (2015) A case for agent-based models in organizational behavior and team research. Team Perform Manag 21(1/2):37–50
Smaldino PE, Calanchini J, Pickett CL (2015) Theory development with agent-based models. Organ Psychol Rev 5:300–317
Ten Broeke G, Van Voorn G, Ligtenberg A (2016) Which sensitivity analysis method should I use for my agent-based model? J Artif Soc Soc Simul 19:5
Tracy M, Cerdá M, Keyes KM (2018) Agent-based modeling in public health: current applications and future directions. Annu Rev Public Health 39:77–94
Von Neumann J (1951) The general and logical theory of automata. In: Jeffress LA (ed) Cerebral mechanisms in behavior: the Hixon symposium. Wiley, New York, pp 1–14
Von Neumann J (1966) Theory of self-reproducing automata. Edited and completed by Arthur W. Burks. University of Illinois Press, Urbana
Wall F (2015) Agent-based modeling in managerial science: an illustrative survey and study. Rev Manag Sci 10:135–193
Wang C, Tang Y, Kassem MA, Li H, Zhizhan W (2021) Fire evacuation visualization in nursing homes based on agent and cellular automata. J Saf Sci Resil 2(4):181–198

Wilensky U (1997) NetLogo traffic basic model. Northwestern University Center for Connected Learning and Computer-Based Modeling, Evanston

Wilensky U (2023) NetLogo user manual version 6.4.0 November 15, 2023. Northwestern University Center for Connected Learning and Computer-Based Modeling, Evanston. https://ccl.northwestern.edu/netlogo/docs. Accessed 8 June 2024

Wilensky U, Rand W (2015) An introduction to agent-based modeling. MIT Press, Cambridge, MA

Yang Y, Deng J, Xie C-c, Jiang Y-t (2014) Design and implementation of fire safety evacuation simulation software based on cellular automata model. Procedia Eng 71:364–371

Zhang H, Vorobeychik Y (2019) Empirically grounded agent-based models of innovation diffusion: a critical review. Artif Intell Rev 52:707–741

Infecting

Keywords Compartmental model · Mask wearing · Nonpharmaceutical interventions · Pandemic · Susceptible-infectious-recovered (SIR) model · To Mask Or Not To Mask? ABM

1 Introduction

When the global pandemic caused by coronavirus SARS-CoV-2, better known as Covid-19, erupted in early 2020, a vaccine would not be authorized on an emergency basis for almost a year. Only nonpharmaceutical interventions (NPIs)—lockdowns, quarantines, social distancing, washing hands, wearing masks, restricting travel—could hope to slow the spread of infections, hospitalizations, and deaths. A few small, mostly island nations—Iceland, New Zealand, Singapore, South Korea, Taiwan—successfully tested, traced, and isolated enough people to prevent uncontrolled Covid-19 surges from overwhelming their healthcare systems (Wilson 2020). Many large nations failed to act assertively and suffered severely, with the United States, Brazil, India, Russia, and Mexico experiencing the highest total fatalities. American medical authorities gave contradictory advice about wearing masks in public. Worried about a shortage of N-95 respirators for front-line healthcare workers, the US Surgeon General tweeted on February 29, 2020, "Seriously, people – STOP BUYING MASKS!" On March 8, Dr. Anthony Fauci, National Institute of Allergy and Infectious Diseases and chief medical advisor to the president, said, "When you're in the middle of an outbreak, wearing a mask might make people feel a little better and it might even block a droplet, but it's not providing the perfect

The original version of the chapter has been revised. A correction to this chapter can be found at https://doi.org/10.1007/978-3-031-86199-4_10

Supplementary Information The online version contains supplementary material available at https://doi.org/10.1007/978-3-031-86199-4_3.

D. Knoke, *Network Collective Action*, Lecture Notes in Social Networks,
https://doi.org/10.1007/978-3-031-86199-4_3

protection that people think it is" (Wright 2021: 152–153). On April 3, after a meta-analysis of 172 studies spanning 16 nations revealed that mask wearing reduced the rate of infections, the Center for Disease Control issued new recommendations. Everyone "should wear a cloth face cover when they go out in public, for example to the grocery store or to pick up other necessities" (Yan 2020). The masks were primarily intended to protect other people in case the wearer was infectious but not showing symptoms.

Politicization of masking recommendations during the US presidential election campaign exacerbated a fraught situation. President Donald Trump constantly spread disinformation, pushing conspiratorial theories about the Chinese origins of the coronavirus, promoting dangerous "cures" by injecting bleach and sunlight into the body, and endorsing anti-malarial and animal deworming medicines. His grandiose narcissistic personality overrode his lifelong germophobic dread of contamination:

> Maskless, he appeared defiant. Masculine, invulnerable, whereas to wear a mask would be caving in, being weak; it might "send the wrong message" and hurt his chances for reelection. Eventually he would come to believe that people wore masks to show their disapproval of him. And yet, from his perspective, exposing himself to microbes that could readily kill him must have seemed heroic (Wright 2021: 151–152)

Republican governors who dared to mandate masks risked ridicule by their national leader and the loss of votes from Covid-19 denialists in their party. Following the CDC's April recommendations, state governors began either to mandate mask-wearing in most public places, to issue less-restrictive mandates, or to impose no mandates at all. The most important predictor was political party, with Republican governors delaying statewide indoor mask mandates by an average 98 days. "This finding highlights a key challenge to public efforts to increase mask wearing, one of the most effective tools for preventing the spread … while restoring economic activity" (Adolph et al. 2022: 24). Eleven states never imposed mandates and some blocked local governments from doing so (Markowitz 2023). By the end of 2020, the 11 states without masking mandates were all led by Republican governors and all had given their electoral college votes for Trump in the 2016 presidential (Adolph et al. 2022).

What evidence showed that face mask mandates actually reduced the severity of Covid-19 outbreaks? In April 2020, the COVID States Project began surveying approximately 20,000 people in each of 20 waves across 3 years to explore their support and practices of public health measures. By comparing state-level mortality rates and mask wearing from June 1 to December 31, 2020, researchers concluded that states without mask mandates had 30% higher death rates per 100,000 residents.

> Correlation is not causation. Those states that had high levels of mask wearing generally had more stringent policies, as well as higher levels of social distancing and avoidance of crowds and public places. Further, there are certainly other, pre-existing, factors that influence COVID-related mortality numbers (like age of the population; density of housing, etc.). Pulling apart that causal tangle is extremely difficult. But there is little doubt that mask mandates and mask wearing at the population level were associated with substantially lower levels of mortality. (Lazer et al. 2023)

In contrast, a meta-analysis of 78 studies spanning a wide range of viral epidemics, conducted by the Cochrane Library, reached a more cautious conclusion:

> There is uncertainty about the effects of face masks. The low to moderate certainty of evidence means our confidence in the effect estimate is limited, and that the true effect may be different from the observed estimate of the effect. The pooled results of RCTs [randomised controlled trials] did not show a clear reduction in respiratory viral infection with the use of medical/surgical masks (Jefferson et al. 2023: 3)

That uncertainty did not prevent conservative pundits from drawing their own interpretations that masking mandates are ineffective (Tierney 2023; Stephens 2023). In the subsequent kerfuffle, the editor in chief of the Cochrane Library apologized for the misleading interpretations of its review and said the summary would be revised (Tufekci 2023). The preponderance of evidence indicated that community-wide wearing of face masks, in conjunction with other NPI practices, conferred protection against transmission of Covid-19 infection.

This chapter reviews compartmental epidemiological models, then inspects applications of agent-based models to the Covid-19 pandemic. After examining evidence pandemic partisanship in the US, it presents an agent-based model demonstrating how differential mask-wearing of political partisans results in divergent collective health outcomes. It concludes with a discussion of America's pandemic polarization as a symptom of declining support and growing distrust in institutional norms.

2 Compartmental Epidemiological Models

Compartmental models comprise the earliest epidemiological efforts to predict the course of infectious disease outbreaks in a population (Ross 1916; Kermack and McKendrick 1927; Heesterbeek 2002). People are classified into a small number of discrete categories and average rates of transition between categories are measured empirically. The simplest SIR model sorts a population of N people into three compartments at each interval of time: Susceptible, Infectious, Recovered (also called Removed in versions where infected people can die). Figure 1 is a schematic of a SIR model with fixed rates of transition per time interval from Susceptible to Infectious (at rate β) and from Infectious to Recovered (at rate γ). For some diseases, such as mumps and measles, recovered people usually gain lifelong immunity against reinfection. In other diseases—such as malaria and sexually transmitted infections like chlamydia and gonorrhea—recovered people are susceptible to reinfection and hence the model returns them to the Susceptible compartment (at rate ξ). The basic SIR model assumes no changes in population size occur because of birth, death, or migration. It also assumes homogeneous mixing: interpersonal contacts occur at random throughout the population and do not differ across subgroups.

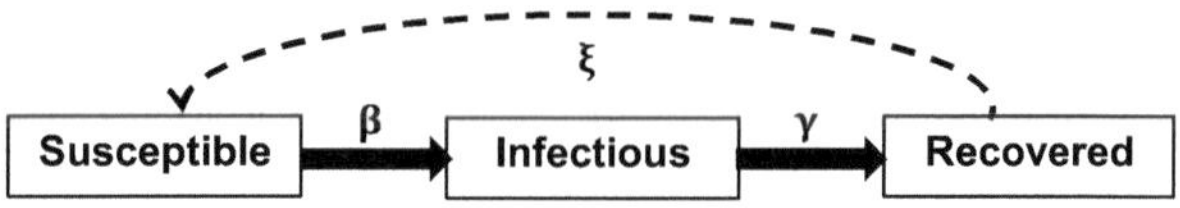

Fig. 1 Simple SIR model of disease

A key factor for forecasting the spread of a disease is its basic reproduction number (R0, pronounced "R naught"), the mean number of people infected by a single infectious person before recovering or removal. If the mean is great than one, the disease grows exponentially (epidemic). If R0 is less than one, the disease dwindles until it dies out. If R0 = 1, then the disease is endemic, spreading at a constant rate in the population. Respiratory diseases differ in their ease of spread among four major mechanisms of transmission (direct physical contact, indirect contact, large droplets, and fine aerosols). Scientists "know little about the relative contribution of each mode of transmission of a particular virus in different settings, and how its variation affects transmissibility and transmission dynamics" (Leung 2021). When a new virus emerges, such as SARS-CoV-2 or its many variants, modeling its likely spread is challenged by many unknown parameters. Early twenty-first century respiratory virus epidemics varied widely in ease of spreading due to differing R0s. Severe Acute Respiratory Syndrome (SARS) and Middle East Respiratory Syndrome (MERS) both had very low R0s (0.58 and 0.69, respectively), which limited their impacts in China and the Middle East to a few thousand cases (Abdelrahman et al. 2020). In contrast, SARS-CoV-2 was extremely contagious and infectious, with an estimated R0 = 3.1 for the Italian outbreak (D'Arienzo and Coniglio 2020); R0 = 3.2 for the Danish Omicron variant (Ito et al. 2022); and a range in the US from R0 = 2.3 for Wyoming to R0 = 7.1 for New Jersey (Mallela et al. 2022).

Basic SIR models can be elaborated to include more compartments and complex transition rates. For example, some diseases have an important latency or incubation period from the time a person becomes infected but is not yet infectious to others. The SEIR model adds an Exposed compartment between Susceptible and Infectious. The time, in days, between Exposed and Infectious varies among diseases. Prior to the availability of a vaccine in 1995, almost every child was infected with chicken pox, which took from 10 to 21 days to develop after exposure. In the year before Covid-19 vaccines were available, the incubation period ranged between 48 h and 14 days, during which people could pass the virus to other before exhibiting symptoms themselves (and many infected people never showed any symptoms). Public health officials urged quarantining and testing after suspected exposure to limit the transmission of Covid-19 during the exposure interval. Other examples of expanded SIR models include compartments for people who are Vaccinated (SIRV) and Deceased (SIRD).

Formulated as a set of ordinary differential equations, compartmental models can simulate epidemic trends based on transition rates estimated from initial field reports (Tolles and Luong 2020). Predictions of the frequencies or proportions of a population in various compartments at future intervals may reveal scenarios where high levels of infection overwhelm a healthcare systems' capacity to treat all seriously ill patients. Models can be rerun using alternative assumptions about the impacts of public health mitigation strategies (quarantines, lockdowns, social distancing, masking) on reducing the R0 reproduction number and slowing the rates of transition among Susceptible, Exposed, and Infectious compartments. "Flattening the curve" is a popular catch-phrase describing efforts to reduce the number of susceptible people so the peak of infections is lower, and the time for the number of infections to fall is shortened.

Model projections are only as good as their assumptions are realistic. During the early weeks of the Covid-19 pandemic, modelers at the Imperial College London predicted that, in the absence of any mitigation and suppression strategies, 81% of the populations of the United States and United Kingdom would be infected, with 2,200,000 deaths in the US and 510,000 UK deaths (Ferguson et al. 2020). They argued that five NPIs—case isolation in the home, voluntary home quarantine, social distancing of those over 70 years of age, social distancing of the entire population, and closure of schools and universities—had "the potential to suppress transmission below the threshold of R = 1 required to rapidly reduce case incidence" (p. 14). Reaction was swift, with other researchers faulting the Imperial College model for bad data and flawed assumptions "driving the world to adopt harsh distancing measures and forget the concept of herd immunity" (Boretti 2020; see also Winsberg et al. 2020; van Basshuysen and White 2021). In the end, the rapid creation of vaccines and their fitful adoption eventually brought the pandemic under control as it edged closer toward the catastrophic Imperial College predictions. On May 5, 2023, the World Health Organization declared an end to the Covid-19 emergency. The WHO dashboard estimated that Covid-19 had cumulatively infected more than 767 million people worldwide, and more than 6,938,000 had died (World Health Organization 2023). Undetected Covid-19 infections may have been 52% of the total number (Melis and Littera 2021), meaning that worldwide infections could have exceeded one billion. The US led all nations with more than 1.1 million deaths (Johns Hopkins University 2023).

3 Covid-19 ABMs

Agent-based modelling of infectious diseases increased markedly in recent years in public health and epidemiological research (Tracy et al. 2018; Bershteyn et al. 2022). An early simulation was the NetLogo program "Virus," written by Uri Wilensky (1998), that tracked the progress of disease. It's initialized by randomly scattering a small number of infectious persons among dozens of susceptibles. As agents wander randomly around their world and come into contact with one another, an infectious person has a random probability of infecting a susceptible, who either recovers with a year of immunity, or dies from disease or from old age. Because people can reproduce, and their newborns are susceptible, the population may grow, up to the ABM world's carrying capacity of 300, or shrink if more people die than are born. A modeler can use sliders to change such parameters as population size, viral transmissibility, and duration of infectiousness. Alternative scenarios can be explored by varying model parameters; for example, reducing the population size (density) to simulate social distancing which lowers the frequency of contact between infectious and susceptible agents.

The Covid-19 pandemic proliferated ABM efforts to predict the pace of viral spread and, in the absence of vaccines in the first year, to investigate the impact of non-pharmaceutical interventions such as social distancing and masking. One bibliometric analysis identified 194 articles with agent-based models of the pandemic

(Tang et al. 2022). That literature is too vast to summarize in detail. Instead, this section briefly summarizes three exemplary models that simulated the impacts of restrictions on the spread of Covid-19 infections. Lindsay Álvarez and Sergio Rojas-Galeano (2020) created a Covid-19 ABM with seven constraints: Social Distancing, Case Isolation, Home Quarantine, Total Lockdown, Sentinel Testing, Masking, and Zonal Enforcement. The latter divides a city into districts where movement between the zones is prohibited or restricted (e.g., curfews). Mask protection parameters varied the chance of contagion when two people encounter at 3% if both wear masks, 8% if an infected person is masked but a susceptible is not, 50% if a susceptible is masked but infected person is not, and 90% whether neither is masked. After repeating every scenario 30 times for a simulated 60-day timeline, the authors concluded that "simultaneous application of the Zone Enforcing intervention with the other NPIs, again yields an improvement on the mitigation impact compared to the single NPI scenarios (more survivors, fewer deaths)" (p. 13). However, they acknowledged that limiting interzonal movements may prove difficult when people live and work in different districts. Two other recommendations are that Sentinel Testing should be given priority, while Mask Protection for everyone "is even a cheaper strategy authorities should strengthen as an alternative to declaring costly lengthy Total Lockdowns" (p. 14).

Dale Brearcliffe (2020) focused his agent-based SEIR model on the efficacy of homemade masks compared to N95, medical, and no mask wearing. "Collective use of homemade masks in the simulation demonstrated a positive difference in high adoption scenarios." Simulations steadily increased the homemade mask adoption rate from 0% to 95% in increments of 5%. Based on 500 model runs for each of the 20 levels of adoption, he found that increasing collective use of homemade masks monotonically reduced that number of exposed and infected agents each day, pushing the peak infection day further from start of the pandemic, thereby preventing a healthcare system from becoming overwhelmed by cases too numerous to handle. The mean day when the greatest number of people were affected reached its longest delay when 85% of the population wore homemade masks, then decreased after that. "The model demonstrates that at high levels of adoption even a mix of questionable quality homemade masks can 'flatten the curve' without immediate, sever [sic] economic cost of staying at home." Unfortunately, Brearcliffe concluded, a "goal of herd immunity by non-pharmaceutical means … takes place only at levels of adoption that may not be possible to achieve" (p. 8).

Covasim (COVID-19 Agent-Based Simulator) is an open-source, high-performance, general purpose ABM designed to be "capable of informing real-world policy decisions for a variety of national and subnational settings" (Kerr et al. 2021: 3). The model calculates disease transmission as infected agents move through their daily contact networks in key settings: households, schools, workplaces, and the general community. Covasim comes preloaded with a country's demographic data, such as age distributions and household sizes, and is integrated with SynthPops, "an open-source data-driven model capable of generating realistic synthetic contact networks for populations" (p. 9). The model has been used by researchers and public health officials in more than a dozen nations to model the effects of alternative interventions on disease transmission; for example, reductions in transmissibility per contact through mask wearing and maintain physical distance versus complete school and workplace closures.

4 Pandemic Partisanship

Just months into the US Covid-19 outbreak, large differences emerged in the mask-wearing attitudes and behaviors of Republican and Democratic partisans. Partisan disagreements about public health policies grew rapidly during the pandemic, exacerbated by distrust of public institutions in general. "[S]tarting around the time President Trump revealed his own COVID-19 diagnosis, the correspondence between partisanship and trust itself began to increase, with Republicans expressing less trust and Democrats reporting greater trust" (Hegland et al. 2022: 70). The covariation of partisanship and distrust of public health institutions deepened the gulf in acceptance or rejection of non-pharmaceutical interventions and in subsequently in rates of vaccination.

Polarization was evident in public opinion polling during the first year of the pandemic when vaccines were unavailable. A national Gallup tracking poll ask respondents whether they had "worn a mask on your face when outside your home?" In April 2020, 77% of Democrats and Democratic leaners and 53% of Republicans and Republican leaners said yes; by June, the gap had widened in 97% of Democrats and 68% of Republicans (Makridis and Rothwell 2020: 59). A Gallup poll in July found a minority of Republicans responding that they "always" (24%) or "very often" (22%) wore masks outside the home in contrast to a large majority of Democrats (61% "always", 33% "very often") (Brenan 2020). A Pew survey in June found that 76% of Democrats reporting wearing a mask in stores or other businesses "all or most the time in the past month" compared to 53% of Republicans. A month later the masking differential had narrowed to 92% of Democrats and 75% of Republicans (Kramer 2020: 2; see also van Kessel and Quinn 2020). A series of Pew polls (Deane et al. 2021) uncovered other divergent partisan pandemic opinions, including: perception of the outbreak as a major threat (82% Democrats, 41% Republicans in February 2021); belief that news media coverage was largely accurate (66% Democrats, 31% Republicans in April 2020); and "requiring most businesses other than grocery stores and pharmacies to close" (81% Democrats, 61% Republicans in March 2020). America stood out in a 14-nation survey as the most politically divided over its governmental handling of the pandemic (Dimock and Wike 2020). The concluding section of this chapter discusses some origins of the politically polarized American pandemic experience.

If lower rates of mask wearing increase the risk of coronavirus infection, one consequence would be higher mortality among Republicans than Democrats. A county-level analysis of Covid-19 deaths during the first pandemic year found that "Republican leaning counties with loose mask mandates experienced up to 3 times higher death rates than Democrat-leaning counties, after controlling multiple social vulnerability measures" (Kaashoek et al. 2022: 1). Partisanship was measured by a county's lean toward the Democratic or Republican 2020 presidential candidate (Biden or Trump) and the state governor's party affiliation. In the most highly polarized counties, "the median death rate of counties with strongest Republican political leaning is between 40% and 300% greater than the median death of counties with the strongest Democratic political leaning depending on the stringency of governor interventions. ... Even after controlling for a diverse array of social vulnerabilities, the importance of political leaning in predicting death rate either doubled or tripled depending on the type of model from period 2 to period 3" (pp. 6–7). The monthly trend from February to December 2020 saw initially higher Democratic rates reversed, with Republican county rates surpassing Democratic county rates by the end of the study period (Chen and Karim 2022). Political differences in geographic mortality patterns arose long before the coronavirus outbreak (Denworth 2022), and disentangling long-term trends from the impact of masking mandate policies would require fine-grained data that were unavailable. The partisan polarization in masking behavior described above inspired the agent-based model in the next section.

5 To Mask Or Not To Mask? ABM

"To Mask Or Not To Mask?" is a NetLogo agent-based model created to demonstrate the health consequences of partisan differences in mask wearing during a pandemic. The political party identifiers are either Republicans or Democrats; Independents are excluded to simplify the analysis. As shown in Fig. 1, the sliders on the left side allow an investigator to set the number of agents (Population_Size) and the percentages in four party-masking combinations. Based on Gallup and Pew polls in June 2020, the default values and agent colors are: 23% mask-wearing Republicans (red), 21% unmasked Republicans (pink), 43% mask-wearing Democrats (blue), and 13% unmasking Democrats (cyan). The distribution can be changed to investigate other political conditions, for example, a red state with a larger Republican population and lower levels of masking wearing. Researchers should take care that the percentages on the four sliders always add up to 100%.

The three sliders on the lower left side determine the outcomes of encounters as the agents wander randomly through their world. At the start, 1% of the population is randomly selected to be infected and turn black; e.g., if the population is 1000, ten agents are infected. When an infected agent enters a patch occupied by an susceptible agent, the chance of infection depends on one of the two Infectiousness sliders. If a random number between 1 and 100 is below the number on the slider, the virus is transmitted and the susceptible agent turns black. To contrast the consequences of masking-wearing, the default transmission rates in this experiment were set at 10% for mask-wearers of both parties and 90% for the unmasked members of both parties. If the two Infectiousness rates were to be set at the same value, masking status would have no differential impact on viral transmission.

Finally, the Removal_Rate slider controls how long an infected agent remains infectious or turns green and stops moving. When a random number from 1 to 100 is less than the rate set on the slider, an agent is removed. In Fig. 2, the default removal rate is set at 4, which gives an infected agent a 3% chance of removal per unit of time (i.e., 3% of the random numbers are less than 4). The simulation halts when all infected agents have been removed, that is, no black agents remain. Because this ABM emphasizes the transition from susceptible to infectious, it does not distinguish whether a removed agent died, was quarantined or hospitalized, recovered with immunity, or returned to susceptibility for possible future reinfection.

Figure 2 is a screen capture of the end results for a single run of the model, taken at tick 175, when a large number of agents had already been infected and removed. A small number of still-infected agents continued to spread the virus until tick 399, when the last infected agent was removed. Figure 3 shows a plotter and seven reporters that track the S-shaped curves of the four groups. Because viruses don't care about people's party affiliations, no substantial differences occur between the Republicans and Democrats who wear masks, nor between the unmasked Republicans and Democrats. However, the percent infected among the unmasked accelerated rapidly relative to the slower increase among the mask-wearers. Because almost half the Republicans didn't wear masks, compared about one-fourth of Democrats, the top three monitors reveal that at the end a majority of Republicans

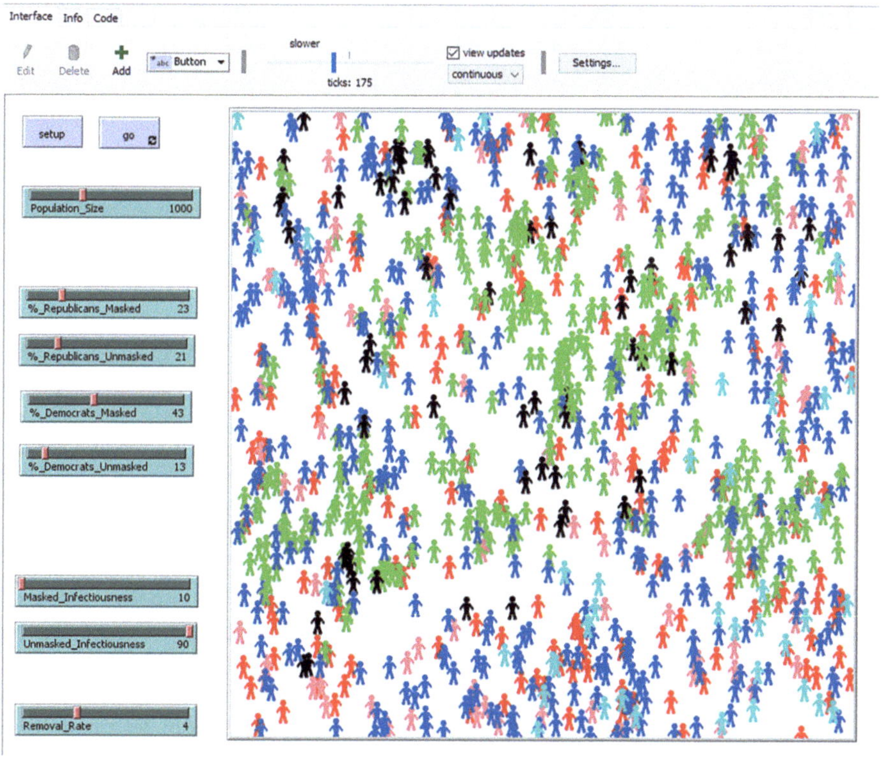

Fig. 2 Sliders and graphics window at 175 ticks

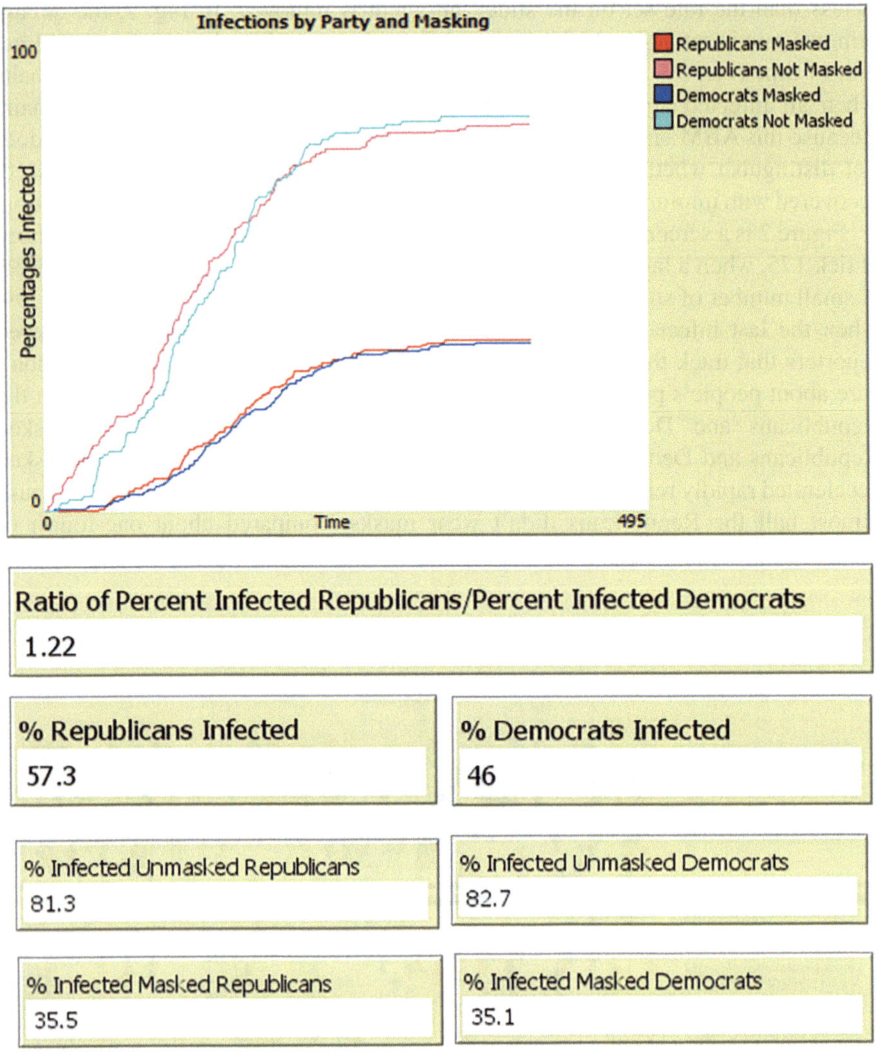

Fig. 3 Plot and reporters when run ended at 399 ticks

had become infected and the ratio of infected Republicans to infected Democrats was 1.22 (i.e., 57.3% divided by 46.0% = 22% larger).

Using NetLogo's BehaviorSpace software, I conducted two experiments, each with 200 repetitions. All sliders were fixed at the values in Fig. 1, except that the Removal_Rate slider was set to 3 in the first experiment and set to 5 in the second experiment. Table 1 summarizes the results of the two experiment. In the first, the lower removal rate meant that infected agents had longer durations to transmit the virus to others whom they encounter. On average, the 200 repetitions ended at 456 ticks. As a result, well more than half the identifiers of both parties were infected

Table 1 Two experiments with different removal rates

	Experiment 1	Experiment 2
Removal rate	3	5
Mean % republicans infected	72.2	36.2
Mean % democrats infected	61.0	26.8
Mean ratio reps/dems	1.17	1.31
Standard deviation	0.05	0.13
Median ratio reps/dems	1.16	1.30
Median duration (ticks)	456	314
Number of repetitions	200	200

and the partisan differential was only 1.17 higher for Republicans. In the second experiment, the faster removal rate meant the runs ended sooner, at 314 ticks on average; that much less than half the members of each party were infected; but, the partisan differential was almost twice as large, 1.31 higher for Republicans.

An important inference of the experimental outcomes is that infected people must be removed from circulation as rapidly as possible. That desideratum was especially crucial during the Covid-19 pandemic because in many cases the latency period (time interval between being infected and becoming infectious) was shorter than its incubation period (interval from infection to the appearance of clinical symptoms) (Xin et al. 2022). In other cases, people with asymptomatic infections unknowingly passed the disease to susceptible people. For example, one meta-analysis estimated an asymptomatic transmission rate of 24.51% (Ravindra et al. 2022). Unfortunately, tracking, testing, and isolating the infected before they could wreak havoc on their friends and neighbors proved beyond the capability of most public healthcare systems in the world.

6 Discussion

The assumption of random mixing in "To Mask Or Not To Mask" is undoubtedly wrong, as stratification researchers and social network analysts regularly demonstrate that people prefer to interact with others having similar class, education, income, race, age, gender, politics, and other attributes. To answer George Box's question—how wrong a model has to be to not be useful—depends on whether the ABM offers insights into the socio-political dynamics of the pandemic. "To Mask Or Not To Mask" was not intended to reproduce in intricate detail the infection trajectories during the first year of the US Covid-19 outbreak, nor was it meant to predict future pandemic SIR patterns. The model was meant to assess whether a few simple assumptions about differential partisan masking behaviors and viral transmissibility, informed by national mask-wearing data collected near the beginning of the pandemic, could explain the tragic collective outcomes. Across repeated experiments with varied parameters, the recurrent results were that higher proportions of Republicans than Democrats succumbed to infections. On that basis, I think "To Mask Or Not To Mask" may qualify as a useful model.

Pandemic political polarization was just one symptom of the ongoing decline over the past half century in American democratic norms and values, growing disbelief in science and medicine, and rising distrust in institutional authorities. The widening gulf between adherents of the two political parties is rooted in the "powerful alignment of ethnicity, ideology, and religion on each side of the divide – what we call the 'iron triangle' of U.S. polarization" (Carothers and O'Donohue 2019). Fanners of the flames of ideological hostility are political party elites, social movement activists, and media platforms that caricature opponents as enemies (e.g., Shin et al. 2022; Young et al. 2022). In this toxic brew, refusing to wear a face mask symbolizes resistance to governmental restrictions on personal liberties, on a par with resistance to gun control and affirmative action. Political polarization widens when society is fragmented into densely interconnected subgroups and weak or absent intergroup connections, creating echo chambers and filter bubbles (Barberá 2020; Wang and Qian 2021; Interian et al. 2022; Olteniceanu et al. 2022). People falling into online echo chambers receive only filtered information which they already believe, thereby reinforcing their confirmation biases (Al Atiqi 2023; Jiang et al. 2021; Recuero et al. 2021; Villa et al. 2021). Men were less willing than women to don masks (Capraro and Barcelo 2020; Brenan 2020), perhaps from fear of appearing weak and unmasculine, while ironically the coronavirus killed males at a higher rate than females.

References

Abdelrahman Z, Li M, Wang X (2020) Comparative review of SARS-CoV-2, SARS-CoV, MERS-CoV, and Influenza A respiratory viruses. Front Immunol 11:552909

Adolph C, Amano K, Bang-Jensen B, Fullman N, Magistro B, Reinke G, Wilkerson J (2022) Governor partisanship explains the adoption of statewide mask mandates in response to COVID-19. State Polit Policy Q 22:24–49

Al Atiqi M (2023) Simulating Echo chamber and polarization problems in social media. In: Echo chamber and polarization in social media: an agent-based modeling approach. Springer Nature, Singapore, pp 17–40

Álvarez L, Rojas-Galeano S (2020) Simulation of non-pharmaceutical interventions on COVID-19 with an agent-based modal of zonal restraint. medRxiv 20130542. www.medrxiv.org/content/10.1101/2020.06.13.20130542v1.full. Accessed 29 Nov 2024

Barberá P (2020) Social media, echo chambers, and political polarization. In: Persily N, Tucker JA (eds) Social media and democracy: the state of the field and prospects for reform. Cambridge University Press, New York, pp 34–55

Bershteyn A, Kim H-Y, Scott Braithwaite R (2022) Real-time infectious disease modeling to inform emergency public health decision making. Annu Rev Public Health 43:397–418

Boretti A (2020) After less than 2 months, the simulations that drove the world to strict lockdown appear to be wrong, the same of the policies they generated. Health Serv Res Manag Epidemiol 7:1–11

Brearcliffe D (2020) Non-pharmaceutical herd immunity using homemade masks. Paper presented to International conference on social computing, behavioral-cultural modeling & prediction and behavior representation in modeling and simulation, 20 October. www.tangledinfo.com/sites/default/files/2021-07/Non-Pharmaceutical%20Herd%20Immunity%20using%20Homemade%20Masks-Preprint.pdf. Accessed 1 June 2023

Brenan M (2020) Americans' face mask usage varies greatly by demographics. Gallup News. https://news.gallup.com/poll/315590/americans-face-mask-usage-varies-greatly-demographics.aspx. Accessed 2 June 2023

Capraro V, Barcelo H (2020) The effect of messaging and gender on intentions to wear a face covering to slow down COVID-19 transmission. arXiv:200505467. https://psyarxiv.com/tg7vz. Accessed 2 June 2023

Carothers T, O'Donohue A (2019) How to understand the global spread of political polarization. Carnegie Endowment for International Peace, 1 October. https://carnegieendowment.org/2019/10/01/how-to-understand-global-spread-of-political-polarization-pub-79893. Accessed 2 June 2023

Chen H-F, Karim SA (2022) Relationship between political partisanship and COVID-19 deaths: future implications for public health. J Public Health 44:716–723

D'Arienzo M, Coniglio A (2020) Assessment of the SARS-CoV-2 basic reproduction number, R0, based on the early phase of COVID-19 outbreak in Italy. Biosaf Health 2(2):57–59

Deane C, Parker K, Gramlich J (2021) A year of U.S. public opinion on the coronavirus pandemic. Pew Research Center, 5 March. www.pewresearch.org/2021/03/05/a-year-of-u-s-public-opinion-on-the-coronavirus-pandemic. Accessed 19 May 2022

Denworth L (2022) People in republican counties have higher death rates than those in democratic counties. Sci Am

Dimock M, Wike R (2020) America is exceptional in the nature of its political divide. Pew Research Center, 13 November. www.pewresearch.org/short-reads/2020/11/13/america-is-exceptional-in-the-nature-of-its-political-divide. Accessed 1 June 2023

Ferguson NM, Laydon D, Nedjati-Gilani G et al (2020) Impact of Non-Pharmaceutical Interventions (NPIs) to reduced COVID-19 mortality and healthcare demand. Imperial College London, London

Heesterbeek JAP (2002) A brief history of R0 and a recipe for its calculation. Acta Biotheor 50(3):189–204

Hegland A, Zhang AL, Zichettella B, Pasek J (2022) A partisan pandemic: how COVID-19 was primed for polarization. Ann Am Acad Polit Soc Sci 700:55–72

Interian R, Marzo RG, Mendoza I, Ribeiro CC (2022) Network polarization, filter bubbles, and echo chambers: an annotated review of measures and reduction methods. Int Trans Oper Res:1–37

Ito K, Piantham C, Nishiura H (2022) Relative instantaneous reproduction number of Omicron SARS-CoV-2 variant with respect to the Delta variant in Denmark. J Med Virol 94:2265–2268

Jefferson T, Dooley L, Ferroni E, Al-Ansary LA, van Driel ML, Bawazeer GA, Jones MA, Hoffmann TC, Clark J, Beller EM, Glasziou PP, Conly JM (2023) Physical interventions to interrupt or reduce the spread of respiratory viruses. Cochrane Database of Systemic Reviews. Issue 1. Art. No. CD006207. www.cochranelibrary.com/cdsr/doi/10.1002/14651858.CD006207.pub6/pdf/full. Accessed 29 Nov 2024

Jiang J, Ren X, Ferrara E (2021) Social media polarization and echo chambers in the context of COVID-19: case study. J Med Internet Res 2(3):e29570

Johns Hopkins University (2023) Coronavirus resource center, Baltimore. https://coronavirus.jhu.edu/map.html. Accessed 22 May 2023

Kaashoek J, Testa C, Chen JT, Stolerman LM, Krieger N, Hanage WP, Santillana M (2022) The evolving roles of US political partisanship and social vulnerability in the COVID-19 pandemic from February 2020–February 2021. PLOS Global Public Health 2(12):e0000557

Kermack WO, McKendrick AG (1927) A contribution to the mathematical theory of epidemics. Proc R Soc Lond Ser A-Contain Pap Math Phys Character 115(772):700–721

Kerr CC, Stuart RM, Mistry D et al (2021) Covasim: an agent-based model of COVID-19 dynamics and interventions. PLoS Comput Biol 17:e1009149

Kramer S (2020) More Americans say they are regularly wearing masks in stores and other businesses. Pew Research Center, 27 August

Lazer D, Baum M, Qu H (2023) Did states with mask mandates have fewer deaths from Covid? The Covid States Project, 27February. www.covidstates.org/blog/did-mask-mandates-reduce-covid-deaths. Accessed 1 June 2023

Leung NHL (2021) Transmissibility and transmission of respiratory viruses. Nat Rev Microbiol 19:528–545

Makridis C, Rothwell JT (2020) The real cost of political polarization: evidence from the COVID-19 pandemic. Covid Econ 34:50–87

Mallela A, Neumann J, Miller EF, Chen Y, Posner RG, Lin YT, Hlavacek WS (2022) Bayesian inference of state-level COVID-19 basic reproduction numbers across the United States. Viruses 14:157

Markowitz A (2023) State-by-state guide to face mask requirements. AARP, 17 May

Melis M, Littera R (2021) Undetected infectives in the Covid-19 pandemic. Int J Infect Dis 104:262–268

Olteniceanu A-M, Van Bevervoorde ME, Flos M, Torlaini R (2022) The network science of echo chambers and why it matters. The Network Pages, June 17. www.networkpages.nl/the-network-science-of-echo... Accessed 30 Nov 2024

Ravindra K, Malik VS, Padhi BK, Goel S, Gupta M (2022) Asymptomatic infection and transmission of COVID-19 among clusters: systematic review and meta-analysis. Public Health 203:100–109

Recuero R, Soares FB, Zago G (2021) Polarization, hyperpartisanship, and echo chambers: how the disinformation about COVID-19 circulates on Twitter. Contracampo-Brazilian J Commun 40:1–16

Ross R (1916) An application of the theory of probabilities to the study of a priori pathometry—part I. Proc R Soc Lond Ser A-Contain Pap Math Phys Character 92(638):204–230

Shin J, Yang A, Liu W, Kim HM, Zhou A, Sun J (2022) Mask-wearing as a partisan issue: social identity and communication of party norms on social media among political elites. So Media Soc 8:1–13

Stephens B (2023) The mask mandates did nothing. Will any lessons be learned? New York Times, 21 February

Tang J, Vinayavekhin S, Weeramongkolkul M, Suksanon C, Pattarapremcharoen K, Thiwathittayanuphap S, Leelawat N (2022) Agent-based simulation and modeling of COVID-19 pandemic: a bibliometric analysis. J Disaster Res 17:93–102

Tierney J (2023) Approximately zero. City Journal, 17 February

Tolles J, Luong TB (2020) Modeling epidemics with compartmental models. JAMA 323(24):2515–2516

Tracy M, Cerdá M, Keyes KM (2018) Agent-based modeling in public health: current applications and future directions. Annu Rev Public Health 39:77–94

Tufekci Z (2023) Here's why the science is clear that masks work. New York Times, 10 March

van Basshuysen P, White L (2021) Were lockdowns justified? A return to the facts and evidence. Kennedy Inst Ethics J 31:405–428

van Kessel P, Quinn D (2020) Both republicans and democrats cite masks as a negative effect of COVID-19, but for very different reasons. Pew Research Center, 29 October

Villa G, Pasi G, Viviani M (2021) Echo chamber detection and analysis: a topology- and content-based approach in the COVID-19 scenario. Soc Netw Anal Min 11:78–94

Wang D, Qian Y (2021) Echo chamber effect in rumor rebuttal discussions about COVID-19 in China: social media content and network analysis study. J Med Internet Res 23(3):e27009

Wilensky U (1998) NetLogo Virus Model. Northwestern University Center for Connected Learning and Computer-Based Modeling, Evanston. http://ccl.northwestern.edu/netlogo/models/Virus. Accessed 1 June 2023

Wilson A (2020) The countries that are succeeding at flattening the curve. Foreign Policy, 2 April. https://foreignpolicy.com/2020/04/02/countries-succeeding-flattening-curve-coronavirus-testing-quarantine. Accessed 1 June 2023

Winsberg EJ, Brennan CW, Surprenant. (2020) How government leaders violated their epistemic duties during the SARS-CoV-2 crisis. Kennedy Inst Ethics J 30:215–242
World Health Organization (2023) WHO coronavirus (COVID-19) dashboard. www.covid19.who.int. Accessed 1 June 2023
Wright L (2021) The plague year: America in the time of Covid. Knopf, New York
Xin H, Li Y, Peng W, Li Z, Lau EHY, Qin Y, Wang L, Cowling BJ, Tsang TK, Li Z (2022) Estimating the latent period of coronavirus disease 2019 (COVID-19). Clin Infect Dis 74:1678–1681
Yan H (2020) Want to prevent another shutdown, save 33,000 lives and protect yourself? Wear a face mask, doctors say. CNN Health, 29 June. www.cnn.com/2020/06/25/health/face-mask-guidance-covid-19/index.html. Accessed 1 June 2023
Young DG, Rasheed H, Bleakley A, Langbaum JB (2022) The politics of mask-wearing: political preferences, reactance, and conflict aversion during COVID. Soc Sci Med 298:114836

Crushing

Keywords Crowd collapse · Crowd management · Crowd safety · Fleeing From Fire ABM · Overcrowding · Stampede · Swiss Cheese Model

1 Introduction

On October 29, 2022, the first Halloween after South Korea lifted its Covid-19 restrictions and social distancing, a crowd of mostly young partygoers wearing costumes and masks swarmed into Itaewon, Seoul's popular, neon-lit nightlife district. Many revelers tried to escape the densely packed main drag, World Food Street, by turning onto a narrow alley running downhill at the rear of bars, nightclubs, and a hotel.

> "I was there, I was fighting to get out because it was so unpleasant and dangerous," said Anthony Spaeth, an editor at the *Korea JoongAng Daily*. "I thought, well, once we get out of the main river of people and turn right into that very short alley which empties out into this big broad street, then this will be over. And when we turned right into the alley, it wasn't over; it got worse. ... The problem was the opposing crowd fighting to get in." (Kwon et al. 2022)

As the alley filled up, hundreds of merrymakers were soon crammed shoulder-to-shoulder. People struggling to return to World Food Street faced an uphill battle against pedestrians surging down into the alley.

At 6:34 p.m., the first of 11 calls to the police alerted them to a looming disaster. "That alley is really dangerous right now people going up and down, so people can't come down, but people keep coming up, it's gonna be crushed. I barely made it to get out but it's too crowded. I think you should control it," the caller said (Slow

The original version of the chapter has been revised. A correction to this chapter can be found at https://doi.org/10.1007/978-3-031-86199-4_10

Supplementary Information The online version contains supplementary material available at https://doi.org/10.1007/978-3-031-86199-4_4.

D. Knoke, *Network Collective Action*, Lecture Notes in Social Networks, https://doi.org/10.1007/978-3-031-86199-4_4

2022). The police promised to respond, but did not dispatch officers. Around 10:00 p.m., people at the top of the slope fell over, setting off a domino effect as bodies toppled onto people trapped below. Unable to inflate their lungs, people who were crushed against walls or beneath others suffocated and died (Lock 2022). When emergency medical crews finally arrived, they counted more than 150 dead, including 26 foreigners, and 133 injured, 37 of them seriously (Jeong et al. 2022). South Korea's police chief later admitted responsibility for poor crowd control planning, with just 137 officers assigned to manage as many as 100,000 people. (For detailed reconstructions of the Itaewon disaster and its causes, see Liang et al. 2024 and Su et al. 2024.) Two years after the Halloween tragedy in Itaewon, a South Korean court sentence the former police station chief to 3 years in prison for occupational negligence in failing to foresee and prevent the man-made disaster. Other police and public officials were handed lighter sentences or acquitted. A bereaved families group condemned the decisions as "absolutely unacceptable" and demanded accountability (Mareca 2024; Yim 2024). The South Korean parliament authorized a new bipartisan investigative committee, while the Seoul metro government installed new safety measures in city areas where Halloween crowds were expected. Whether such actions lead to happier Halloween festivities is uncertain.

Crowd disasters—mass gatherings leading to injuries or deaths—are common occurrences throughout the world, causing approximately 2000 deaths every year. An analysis of press reports from 1900 to 2019 found steadily rising numbers across that time, especially in lower-middle income countries of West Africa and India (Feliciani et al. 2023). This chapter examines how overcrowding in confined spaces can result in crowd collapse and crushing. Just as social distancing is an important strategy for reducing the risk of viral infection, so the risk of crushing can be diminished by preventing overcrowded conditions from arising. After reviewing theoretical explanations of such events, it discusses three agent-based models of real-world crushing incidents. It then presents Fleeing From Fire, an ABM simulating fatal collective behavior in an overcrowded area. The concluding section discusses how crowd management professionals can apply insights from crowd simulation to improve public safety at risky events.

2 Overcrowding

A crowd is a large gathering of people at a physical location. A more precise definition:

> A crowd is a large group of individuals ($N \geq 100$ P) within the same space at the same time whose movements are for a prolonged period of time ($t \geq 60$ s) dependent on predominantly local interactions ($k \geq 1$ P/m^2).
>
> The numbers N (number of individuals), k (density) and t (time) are chosen in a way as to exclude movements during which interaction is non-existent or only present for very short periods of time. (Duives et al. 2013: 194)

Setting the minimum density at one or more people per meter-squared (almost 11 square feet) begs the question, when does overcrowding become potentially injurious and fatal? Two people per square meter can usually move freely without touching one another. At three to four persons, the risk of injury remains low, but a density of five people per square meter has "a high risk of injury from falling and being stepped on" (Pearl 2015: 7). If crowd density reaches or exceeds seven people per square meter "the risk of crowd-related injuries or deaths is extremely high." Once a crowd collapses in on itself, people are trapped so tightly they lose control over their individual movements. Shock waves "can be propagated through the mass sufficient to lift people off their feet and propel distances of 3 m (10 ft) or more" (Fruin 2002). The intense heat of surrounding bodies causes some people to faint standing up. Incapable of falling on the ground to start blood flowing again to the heart and brain, they suffer cardiac arrests. Others, unable to expand their lungs inside chests, die within minutes from compression asphyxiation.

John Fruin (2002) identified two main group motivations for crowd disasters—flight responses and crazes. Perceived or real threats send crowds scurrying to evacuate a confined space through the nearest exits.

> Frequently mislabeled a panic, closer investigation usually shows that flight was a reasonable group reaction under the perceived circumstances. These incidents often show mutual cooperation and assistance among individuals within the group, rather than destructive behavior. (Fruin 2002: 4)

In contrast, a craze is a "competitive rush" toward a valued object "created where participation in an event, or viewing of a public personage, is intensively promoted" (p. 4). On December 3, 1979, 11 young people were crushed to death at a sold-out Who concert in Cincinnati's Riverfront Coliseum when thousands of ticketholders at the west entrance pressed forward, squeezing and trampling as they tried to enter the stadium through a bottleneck (Flippo 1980). Religious shrines are periodically scenes of crushing as pilgrims try to get closer to the objects of their adoration. The deadliest incidents occurred at Saudi Arabia's annual Hajj pilgrimage, with 1426 deaths in 1990 at a Mecca tunnel and more than 2400 deaths in 2015 at a Mina stampede (Langewiesche 2018). On October 1, 2022, in Malang, Indonesia, at least 125 people died and 44 were injured after police fired tear gas in a soccer stadium, setting off a panic in which people attempting to flee the arena were trampled and suffocated (Llewellyn and Tan 2022). At a Hindu religious gathering in Uttar Pradesh, India, more than 100 people were crushed to death on July 2, 2024, when they tried to leave the tent through a narrow exit (Hrishikesh 2024). Such incidents provide the empirical bases for physical and social science efforts to model overcrowding outcomes.

The past few decades saw "an unprecedented wave of crowd simulation studies" (Yang et al. 2020: 1), spurred by advances in computer vision and crowd surveillance technologies. Space limitations allow only brief descriptions of the more prominent approaches. The social force model assumes the movements of pedestrians in a crowd are subjected to three forces: taking the shortest path toward a destination; keeping a distance from other pedestrians, walls, and obstacles; and "attractive effects" such as friends and window displays (Helbing and Molnár 1995). The resulting nonlinearly coupled Langevin equations showed the social force model "is capable of describing the self-organization of several observed collective effects of pedestrian behavior very realistically" (p. 4282). A recent extension of the social forces model introduced heterogeneity of pedestrian physiology (such as age, gender, height, weight, vision range) and psychology (e.g., identity, character, temperament, familiarity) to capture more realistic behaviors (Wu et al. 2021). Using parameters estimated from a survey at Tsinghua University library, simulated crowd evacuations agreed well with empirical observations, for example, young students with strong physiques escaping faster than elderly professors.

Physicists and civil engineers analyze crowd behavior as a liquid whose flow can be explained by the coupled, nonlinear, partial differential equations of fluid dynamics (Helbing 1998; Kok et al. 2016). "The remarkable thing about these equations is that they are conformably mappable and as such may be easily solved in simple geometries. Many problems require modification of these basic equations for specific features of the crowd's behavior" (Hughes 2003: 180). However, experimental evidence indicates that, although conventional motion estimation methods fail at high-density, crowd fluid dynamics could estimate crowd motion only at the global level of analysis (Farooq et al. 2020).

3 ABMs of Disasters

Agent-based simulations use a grid to represent room, building, or passageway with various features like exits, windows, walls, and obstacles, which can influence agent actions and collective outcomes. An agent can move onto and off grid cells, thereby altering the actor's local neighborhood and its proximity to other agents. Importantly, more than one agent may occupy a cell at the same time, simulating crushing effects. ABM scenarios can generate emergent and self-organizing macro-level properties that are predictable neither from individual agent attributes nor by the rules governing their interactions. This section offers a detailed examination of three ABMs that demonstrate how overcrowding may rapidly develop and lead to catastrophic consequences.

Rhode Island Nightclub Fire More than 400 rock fans crowded the dance floor of *The Station*, a wooden roadhouse in Warwick, Rhode Island, on the night of February 20, 2003. As headliner band Great White launched into their signature song, the tour manager ignited pyrotechnics on stage. Within seconds polyurethane foam insulation burst into flames along the walls and top of the drummer's alcove. Flashover fire raced quickly across the ceiling and dark acrid smoke obscured the exit locations. Egress from the nightclub, which did not have sprinklers, was hindered by crowd surges at the main entrance. Almost all the evacuation activity occurred in the 3 min after ignition. By the time firefighters extinguished the blaze, it had killed 100 people and injured more than 200—the deadliest rock concert fire in US history. Brian Butler, a Cable Network News affiliate photographer, gave this eye-witness report:

> At first, there was no panic. Everybody just kind of turned. Most people still just stood there. In the other rooms, the smoke hadn't gotten to them, the flame wasn't that bad, they didn't think anything of it. Well, I guess once we all started to turn toward the door, and we

> got bottlenecked into the front door, people just kept pushing, and eventually everyone popped out of the door, including myself.
>
> That's when I turned back. I went around back. There was no one coming out the back door anymore. I kicked out a side window to try to get people out of there. One guy did crawl out. I went back around the front again, and that's when you saw people stacked on top of each other, trying to get out of the front door. And by then, the black smoke was pouring out over their heads. (CNN Transcripts 2003)

A National Institute of Standards and Technology investigation described the "occupant overload" at the front doors. "Prior to 1-1/2 minutes into the fire, a crowd-crush occurred in the front vestibule which almost entirely disrupted the flow through the main exit. Many people became struck in the prone position in the exterior double doors" (Grosshandler et al. 2005: xx). Among the NIST's recommendations were to increase main entry capacity to accommodate two-thirds of permitted occupant level during an emergency; required staff training and evacuation plans; and improved means for occupants to locate emergency routes (p. xxiii).

PrioritEvac, a free and open source NetLogo program created by Eileen Young and Benigno Aguirre (2021), was designed based on available data culled from post-event interviews and the NIST report. PrioritEvac's environment was constructed from schematics of The Station's floor, window, and door plan. The model scales a patch-length at one-tenth of a meter and set one time-tick equal to 1 s of real time. It incorporates detailed data about people's ages, genders, and locations at the start of the fire. Agent traits include age-specific speed limits; vision impacted by distance, angle, and smoke; and the path taken. The path-finding algorithm "seeks in an ongoing manner the most efficient path from the existing point of an agent to their goal, avoiding fire, smoke, walls and other agents in the simulation. The algorithm runs as soon as an agent determines that they are going to move, and then every tick thereafter until they die or exit the building" (p. 1088). Unlike models that assume agents have no social relationships to one another, PrioritEvac includes group loyalty—staying with "coworkers, friends, dating partners, familial (including married couples), and agents who have multiple kinds of relationships" (p. 1093). Strength of loyalty varies by type of group, with coworkers given the lowest loyalty value and multiple-relations the highest value. The model also takes into account "stigmergic" interactions, indirect coordination via environmental influences, such as people's reactions to observed crowding.

Action begins when agents become alarmed—by fire, smoke, or other nearby agents who are alarmed—and try to escape from the room. They move from patch-to-patch along the route to their chosen exit, avoiding obstacles. Agents who came alone prioritize the goal of personal survival and immediately head for the nearest exit. Those who came in groups set a first goal of searching for their nearest group-member(s). Once the group is in contact, its leader (a club employee, previous visitor, and/or male) determines the nearest exit and the other group members follow-the-leader within two meters. Group loyalty diminishes with proximity of fire or smoke and perception of danger. Sometimes loyalty may fall below a threshold, which can be set by the analyst, and "agents stop prioritizing their group and act as individuals" (p. 1090). Agents are assigned an initial health level—based on a combination of age, gender, and known prior medical condition—which worsens with nearness to fire and smoke, until energy level reaches 0 and the agent dies. The simulation stops when all agents have either died or escaped. The complexity of the simulation requires about 20 hours of computer time per run, a limitation caused primarily by the pathfinding algorithm (p. 1095).

The PrioritEvac model was evaluated against the actual Station fire where 100 people died. The results "indicate that, compared to historical patterns, it reproduces along multiple metrics including a mean of 114 deaths (standard deviation = 38) over 50 runs, which puts the actual result of the fire within one standard deviation of the mean results of the simulation. Overall, the mean differential along all the metrics is 79, significantly outperforming all published ABM models of the Station nightclub fire that did not incorporate social relationships" (p. 1083). The mean number of people who used the main exit or the bar and sunroom windows were also within a standard deviation of the actual outcomes. An additional simulation without the group loyalty component resulted only 45 dead, "demonstrating it is primarily group loyalty and not modeling approach that yielded the overall results" (p. 1094). In other words, people who looked out for others were less likely to survive the inferno.

Shanghai Waterfront Crowd Collapse On December 31, 2014, a deadly crush occurred in Shanghai, near the historic waterfront Bund along the Huangpu River, where a crowd of about 300,000 had gathered to celebrate the New Year. A high-density portion of the crowd collapsed at the staircase connecting Chen Yi Square to the elevated river-view platform, crushing 36 people to death and seriously injuring 49 others (Zhou et al. 2017). The stairs had no separation in the middle to channel people trying to move upwards and downwards. A few minutes before the collapse and crushing, people were constantly colliding at this bottleneck and movement stalled in the middle of the stairs (Liu and Kaneda 2020a: 520).

Based on findings of the official investigation report, physical measures of the site, and a pedestrian-flow survey, Yuanyuan Liu and Toshiyuki Kaneda (2020b) developed an agent-based simulation using Artisoc software. Pedestrian movement was governed by a total of 36 rules, including basic behavior, slow-down, avoidance, and high-density flow. The analysts ran each of six scenarios 20 times, the original Shanghai Bund space layout in 2014 and five alternative designs intended to improve crowd management. The experimental results "revealed a different possible reason for the crowd disaster, which may have been caused by the congestion on top of the staircase instead of the contraction in front of it" (p. 189).

> The high density accumulating on top of the staircase is possibly caused by chaotic flows coming from different directions and will lead to a downward pressure. When the people in front of the stairs can no longer sustain the pressure, someone will ultimately fall and will cause a domino effect. The video recorded by the media confirmed this inference (pp. 186–187)

The alternative spatial designs showed that such changes as capacity reserve and better route planning could increase crowd safety in a high-density environment.

Hajj Stampedes Hajj is a pilgrimage to Makkah, Saudi Arabia, the holiest city for Muslims (Tagliacozzo and Toorawa 2016). It's a religious mandate that must be performed at least once in a lifetime. Annually up to three million Muslim pilgrims (Hajjis) from 140 countries travel to Makkah to engage various rituals within 5 days. Ramy al-Jamarat (stoning the devil) is performed on 3 days when millions of Hajjis walk to the Jamarat area and throw stones at Jamarah pillars. In Tawaf, pilgrims walk seven times counterclockwise around the Kaaba (House of Allah) and stop to kiss the Black Stone mounted on a wall; or, if the crowd is too dense, simply pointing symbolically in its direction. The Say'ee ritual involves walking back and forth seven times in a 400-meter long covered corridor between two hills to the south and north of the Kaaba. A special lane is reserved for disabled persons, such as those walking with canes or being pushed in wheelchairs. Despite safety measures, massive crowd movements periodically lead to overcrowding and crushing with hundreds or thousands of injuries and deaths. For example, 1426 were killed in 1990 at the Jamarat area bottleneck and 2400 died in 2015 at Mina valley (Mahmood et al. 2017). In an effort to improve safety, the Jamarat area was reconstructed in 2010. The older pillars were replaced by larger elliptical pillars and internal columns were eliminated to reduce congestion. Each of the new bridge's five levels serve pedestrians flowing in a single direction so Hajjis would not encounter anyone moving in the opposite direction.

Computer science and security researchers have modelled the crowd dynamics of various Hajj rituals (see Owaidah et al. 2019 for a review). Afnan Alazbah and Bassam Zafar (2019) proposed a model of the Ramy al-Jamarat ritual based on video images from the 2017 Hajj. As pilgrims stopped to throw their stones, crowd flow was slowed or halted, yet people continued arriving from preceding areas, threatening to increase crowd density to dangerous levels. The authors trained human monitors to classify a crowd's status as normal, semi-crowded, or crowded based on camera images without counting heads. Colored traffic lights—green, yellow, red, respectively—were positioned in each of three segments of the Jamarat path. Cameras took images of the segments every few seconds and their monitors changed the lights in the preceding area to signal that pedestrians should stop, slow down, or keep moving. An ABM was implemented with the Any Logic platform containing the Pedestrian Library to simulate Hajji flows in the physical environment. The simulation updated agent locations and speeds at every simulated time step and classified the flow rates of people through the three Jamarat segments as free, constant, crowded, or stampede states "where pedestrians may lose their balance and fall, perhaps getting injured" (p. 11). Without the colored lights, density reached the stampede state, but applying the proposed model successfully kept congestion levels in the normal flow state.

4 Fleeing From Fire ABM

This agent-based model depicts an evacuation from a burning rock concert hall. The screen capture in Fig. 1 shows the stage at the top, with a flames symbolizing fire ignited by band pyrotechnics. Two side doors are located near the stage and the main entrance is at the bottom. When the simulation begins, the randomly scattered blue concert attenders head for one of the three exits: people in the top left quadrant head left, people top right quadrant go right, and the those in the lower half of the room seek to exit through the main entrance. As agents converge on the exits, the density of people occupying a patch may rise to a dangerous level where compression asphyxiation occurs. An analyst can use the slider to set the "DieNumber" anywhere from 2 to 10, meaning that if the patch is occupied with N agents, one of them randomly dies and turns black. If a blue agent reaches a door, it turns green and goes outside. The simulation halts when no agents remain inside the building. Two counters report the percentages escaped and crushed to death at each tick, while the plotter tallies both measures as the simulation progresses. Figure 1 is from a simulation with an audience of 310 people where the DieNumber was set to 3 persons on a patch. After 50 ticks, crowding is already visually evident near the

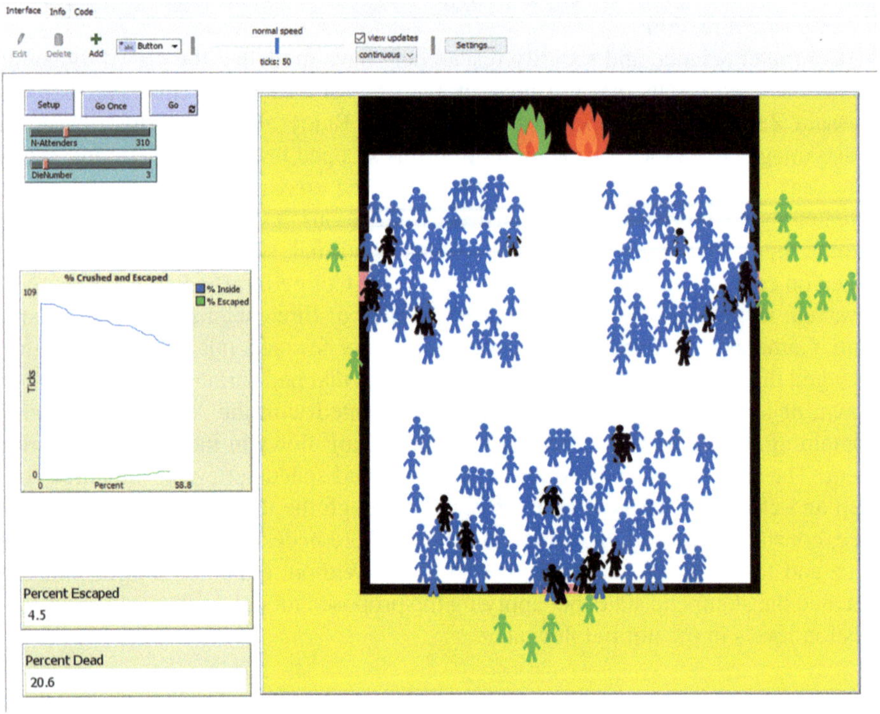

Fig. 1 Fleeing From Fire ABM with 310 attenders at 50 ticks

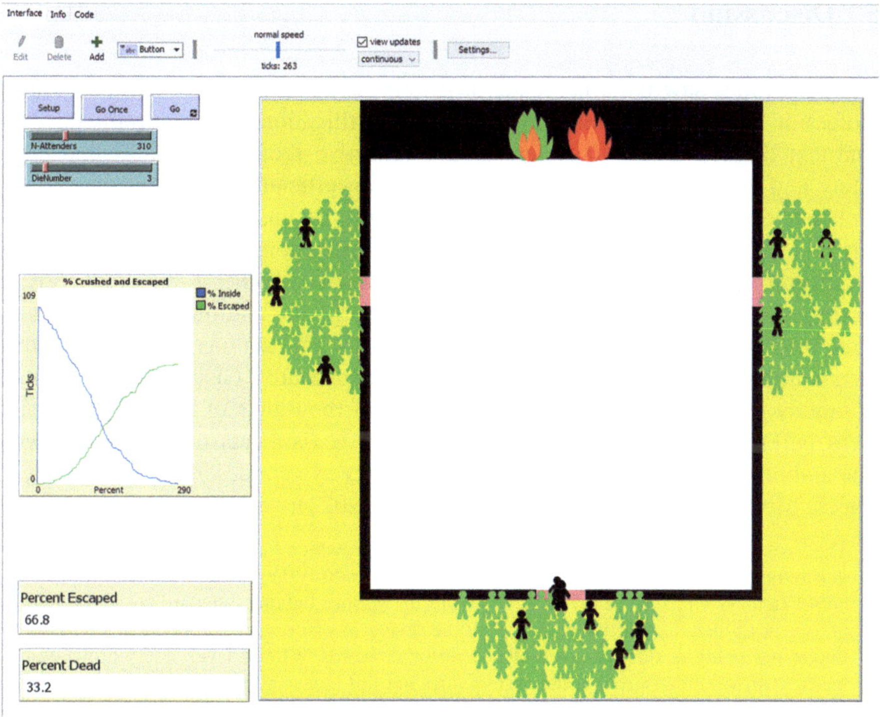

Fig. 2 Fleeing From Fire ABM with 310 attenders at 271 ticks

Table 1 Concert attenders crushed to death and escaped by audience size

	Experimental results	
	Model 4.1	Model 4.2
Mean percent crushed	30.4%	63.1%
Mean percent escaped	69.6%	36.9%
Mean ticks	257.1	250.7
Number of attenders	310	1010
Number of experiments run	200	200

three exits. In Fig. 2 when the run ended after 263 ticks, 33.2% of the attenders had been crushed to death and 66.8% escaped from the burning building.

The number of attenders affects the life-or-death outcomes of this ABM. The N-Attenders slider enables a range of audience sizes from 10 to 1010 by increments of 50. NetLogo's BehaviorSpace software was used to run 200 replications with the N-Attenders slide set at 310 and 1010 agents, respectively, with the DieNumber fixed at 3. Table 1 shows that as crowd size increases, survival rate falls dramatically: 30.4% of the smaller audience were crushed to death and 63.1% of the larger audience was killed.

5 Discussion

Recent decades saw the emergence of crowd management as a multidisciplinary profession with substantial training programs, certification, even university degrees. One text defined crowd management as "Proactive security activities that bring safety and comfort to individuals by facilitating efficient movement of crowds" (Feliciani et al. 2022: 3). The managers responsible for ensuring crowd safety, comfort, and efficiency are employed by a range of institutions and organizations, such as government agencies, transportation companies, weather forecasters, private security, police, fire and ambulance services, hospitals, and morgues. They must share information, plan and schedule events, coordinate actions, anticipate unforeseen contingencies, implement proactive countermeasures. Given the complexities of rapidly changing conditions involving tens of thousands of people in confined areas, building systemic redundancies can reduce the risks catastrophic breakdowns. The authors of a proposed multi-layered of safety system described it as a "Swiss Cheese Model of Crowd Safety" (Haghani et al. 2023):

> The Swiss Cheese Model is named after Swiss cheese, which has a distinctive appearance due to the holes or gaps in its structure. The model visualises these holes as potential weaknesses in a system, and the layers of cheese as the various defences that are put in place to prevent a failure. … But by having multiple layers in place, the likelihood of a critical failure occurring is significantly reduced. In other words, a multi-layer safety protection mechanism would ensure that the system does not fail unless all individual layers fail, an unlikelier event compared to the failure of a single layer. (p. 15)

Crowd modeling by computer simulations is one of the layers for identifying risks and testing alternative crowd management strategies. Because all algorithms have strengths and weaknesses, "selecting a simulator that best satisfies objectives in vital" (Nishinari et al. 2024:241). ABM quality will improve with continual incorporation of data from real-world crowd events.

References

Alazbah A, Zafar B (2019) Pilgrimage (Hajj) crowd management using agent-based method. Int J Found Comput Sci Technol 9:1–17

CNN Transcripts (2003) Breaking news: nightclub fire kills 39 people. Aired, 21 February. https://transcripts.cnn.com/show/bn/date/2003-02-21/segment/09. Accessed 17 Aug 2024

Duives DC, Daamen W, Hoogendoorn SP (2013) State-of-the-art crowd motion simulation models. Transp Res Part C Emerg Technol 37:193–209

Farooq MU, Saad MNBM, Malik AS, Ali YS, Khan SD (2020) Motion estimation of high density crowd using fluid dynamics. Imaging Sci J 68(3):141–155

Feliciani C, Shimura K, Nishinari K (2022) What is crowd management? In: Introduction to crowd management: managing crowds in the digital era: theory and practice. Springer, Cham, pp 1–12

Feliciani C, Corbetta A, Haghani M, Nishinari K (2023) Trends in crowd accidents based on an analysis of press reports. Saf Sci 164:106174

Flippo C (1980) Rock & Roll tragedy: why 11 died at the who's Cincinnati concert. Rolling Stone, 24 January. www.rollingstone.com/feature/rock-roll-tragedy-why-11-died-at-the-whos-cincinnati-concert-93437. Accessed 22 June 2023

Fruin JJ (2002) The causes and prevention of crowd disasters. Originally presented at the First international conference on engineering for crowd safety at London 1993. Revised exclusively for Crowdsafe.com. www.workingwithcrowds.com. Accessed 22 June 2023

Grosshandler W, Bryner N, Madrzykowski D, Kuntz K (2005) Report of the technical investigation of the station nightclub fire (NIST NCSTAR 2), volume 1. National Construction Safety Team Act Reports, National Institute of Standards and Technology, Gaithersburg. https://tsapps.nist.gov/publication/get_pdf.cfm?pub_id=100988. Accessed 17 June 2022

Haghani M, Coughlan M, Crabb B, Dierickx A, Feliciani C, van Gelder R, Geoerg P et al (2023) A roadmap for the future of crowd safety research and practice: introducing the Swiss cheese model of crowd safety and the imperative of a vision zero target. Saf Sci 168:1–22

Helbing D (1998) A fluid dynamic model for the movement of pedestrians. arXiv:cond-mat/9805213v1

Helbing D, Molnár P (1995) Social force model for pedestrian dynamics. Phys Rev E 51:4282–4286

Hrishikesh S (2024) More than 100 killed in crush at India religious event. BBC News. www.bbc.com/news/articles/cv2g7wgj83no. Accessed 2 July 2024

Hughes RL (2003) The flow of human crowds. Annu Rev Fluid Mech 35:169–182

Jeong, Sophie, Gawon Bae, Paula Hancocks, Hilary Whiteman and Jessie Yeung. 2022. "What we know about the deadly Halloween disaster in Seoul." CNN, 1 November. www.cnn.com/2022/10/30/asia/seoul-itaewon-halloween-crush-explainer-intl-hnk/index.html. Accessed 20 June 2023

Kok VJ, Lim MK, Chan CS (2016) Crowd behavior analysis: a review where physics meets biology. Neurocomputing 177:342–362

Kwon J, Montgomery H, Cheung R, Wong A (2022) 'They completely failed': the fatal mistakes that led to South Korea's Halloween tragedy. Vice, 4 November. www.vice.com/en/article/7k8mab/south-korea-itaewon-stampede-halloween-disaster. Accessed 20 June 2023

Langewiesche W (2018) The 10-minute Mecca stampede that made history. Vanity Fair, 9 January. www.vanityfair.com/news/2018/01/the-mecca-stampede-that-made-history-hajj. Accessed 2 July 2024

Liang H, Lee S, Sun J, Wong SC (2024) Unraveling the causes of the Seoul Halloween crowd-crush disaster. PLoS One 19(7):1–21

Liu Y, Kaneda T (2020a) Using agent-based simulation for safety: fact-finding about a crowd accident to improve public space design. Collecti Dyn 5:519–521

Liu Y, Kaneda T (2020b) Using agent-based simulation for public space design based on the Shanghai bund waterfront crowd disaster. Artif Intell Eng Des Anal Manuf 34(2):176–190

Llewellyn A, Tan R (2022) Anger mounts toward police as Indonesia reels from soccer stadium tragedy. Washington Post, 3 October. www.washingtonpost.com/world/2022/10/03/indonesia-soccer-stadium-death-police-gas. Accessed 2 July 2024

Lock S (2022) Crowd crushes: how disasters like Itaewon happen, how they can be prevented, and the 'stampede' myth. The Guardian, 1 November. www.theguardian.com/world/2022/nov/01/how-do-crowd-crushes-happen-stampede-myth-what-happened-in-the-seoul-itaewon-halloween-crush. Accessed 20 June 2023

Mahmood I, Haris M, Sarjoughian H (2017) Analyzing emergency evacuation strategies for mass gatherings using crowd simulation and analysis framework: hajj scenario. In: Proceedings of the 2017 ACM SIGSIM conference on principles of advanced discrete simulation. Association for Computing Machinery, New York, pp 231–240

Mareca T (2024) Two years after deadly Seoul Halloween crush, families still mourn, look for answers. UPI World News, 29 October

Nishinari K, Feliciani C, Jia X, Tanida S (2024) Recent developments in crowd management: theory and applications. J Disaster Res 19:239–247

Owaidah A, Olaru D, Bennamoun M, Sohel F, Khan N (2019) Review of modelling and simulating crowds at mass gathering events: hajj as a case study. J Artif Soc Soc Simul 22. https://jasss.soc.surrey.ac.uk/22/2/9.html. Accessed 4 June 2022

Pearl TH (2015) Crowd crush: how the law leaves American crowds unprotected. Kentucky Law J 104:1–45

Slow O (2022) South Korea: how the Halloween tragedy unfolded. BBC News, 2 November. www.bbc.com/news/world-63448040. Accessed 20 June 2023

Su Z, Cheshmehzangi A, Bentley BL, McDonnell D, Ahmad J, Šegalo S, Da Veiga CP, Xiang Y-T (2024) Deadly yet preventable? Lessons from South Korea's Halloween crowd crush. Disaster Med Public Health Prep 18:1–14

Tagliacozzo E, Toorawa SM (eds) (2016) The hajj: pilgrimage in Islam. Cambridge University Press, New York

Wu W, Chen M, Li J, Liu B, Zheng X (2021) An extended social force model via pedestrian heterogeneity affecting the self-driven force. IEEE Trans Intell Transp Syst:237974–237986

Yang S, Li T, Gong X, Peng B, Jie H (2020) A review on crowd simulation and modeling. Graph Model 111:101081

Yim H (2024) South Korea Halloween crush: police officials jailed over 'man-made' tragedy. Reuters, 30 September

Young E, Aguirre B (2021) PrioritEvac: an Agent-Based Model (ABM) for examining social factors of building fire evacuation. Inf Syst Front 23:1083–1096

Zhou M, Wang M, Zhang J (2017) How are risks generated, developed and amplified? Case study of the crowd collapse at Shanghai bund on 31 December 2014. Int J Disaster Risk Reduct 24:209–215

Rioting

Keywords Bystander · Instigator · Peacekeeper · Riot · Rioters · Sports riot · Threshold model · What a Riot! ABM

1 Introduction

On the evening of June 15, 2011, a downtown Vancouver crowd of 155,000 spectators watched on a big outdoor screen as the hometown Canucks hockey team lost the seventh game of the Stanley Cup finals. During the next 5 h, roughly 1000 rioters threw beer bottles, flipped and burned cars, vandalized food trucks, smashed windows, and looted stores as onlookers cheered. Losses from theft and property damage totaled more than $4 million. Four people were stabbed and 140 others were injured, including nine police officers, and 101 were arrested (Adams 2021). An angry mob of 15 men kicked, punched, and bear-sprayed Robert MacKay, a 39-year-old chef who tried to keep them from breaking into The Bay department store. Bystanders offered the temporarily blinded MacKay water bottles to pour over his head and two men escorted him away from the scene. "MacKay's body was extremely sore, especially his ribs, and he sported a large bump on the back of his head. However, he wasn't seriously injured and opted against going to the hospital. 'I felt like I took some good shots,' he says" (Singh 2021). The Vancouver Police Department's post-riot investigation concluded:

> In short, a riot occurred because hundreds of instigators, many of whom were young, intoxicated males, decided to riot. Another group joined in while the majority of people acted as an enthusiastic audience who encouraged and unwittingly aided the rioters by insulating

The original version of the chapter has been revised. A correction to this chapter can be found at https://doi.org/10.1007/978-3-031-86199-4_10

Supplementary Information The online version contains supplementary material available at https://doi.org/10.1007/978-3-031-86199-4_5.

D. Knoke, *Network Collective Action*, Lecture Notes in Social Networks, https://doi.org/10.1007/978-3-031-86199-4_5

> them from the police by refusing to leave the area. These circumstances were exacerbated by the phenomenon of thousands of attendees using camera phones to record what many seemed to view as "entertainment." (Banel 2021)

In the riot's aftermath, the Vancouver police used facial recognition software on hundreds of photos and videos to identify suspects and prosecutors brought criminal charges against 301 rioters, the vast majority of whom pled guilty. In two trials, five of the six men charged with assaulting Robert MacKay were convicted and sentenced to jail terms up to 8 months.

A worldwide phenomenon, many post-game sports riots seem to erupt spontaneously, propagating without prior planning and on-site coordination by leaders. Evidence is inconclusive whether excessive alcohol consumption on game days contributes to violent fan behavior by lowering their inhibitions against disorderly conduct (Ostrowsky 2018). Less spontaneously, gangs of "football hooligans" arrive expecting to fight during and after matches (Dunning 2000; Zimniak 2020; Andres et al. 2023). Aggressive police attempts to control crowds, such as striking people with batons or shooting them with rubber bullets, may trigger violent fan backlash or panics. An objective of this chapter is to examine an agent-based model of riot initiation, prior to any police intervention, by applying the concept of decision thresholds to a crowd of bystanders.

2 Riot Thresholds

After years of riots in US cities resulted in hundreds of African American deaths, and millions of dollars of property damage, the 1968 Anti-Riot Act (Title 18 U.S. Code Chapter 102 Riots) defined a riot as a public disturbance involving:

> (1) an act or acts of violence by one or more persons part of an assemblage of three or more persons, which act or acts shall constitute a clear and present danger of, or shall result in, damage or injury to the property of any other person or to the person of any other individual or (2) a threat or threats of the commission of an act or acts of violence by one or more persons part of an assemblage of three or more persons having, individually or collectively, the ability of immediate execution of such threat or threats, where the performance of the threatened act or acts of violence would constitute a clear and present danger of, or would result in, damage or injury to the property of any other person or to the person of any other individual.

Examples of riotous acts are inciting others to riot, destruction of property by breaking windows and doors, looting stores, throwing stones and bricks, assaults with poles and machetes, discharging firearms that injure or kill people. Urban riots frequently erupt in marginalized and socioeconomically deprived areas (Holdo and Bengtsson 2020), although sports riots obviously involve fans who can afford to purchase often expensive admission tickets.

Threshold models, sometimes called tipping-point models, are an adaptation of the spread of infectious disease to social collective action. Instead of contagion by viruses or bacteria, an onlooker person perceives the behaviors of others in an assembly, then engages in that behavior if the magnitude of those perceptions reaches or exceeds some level. Individuals have differing thresholds for activation, so someone with a low threshold acts more quickly, while those with higher thresholds activate more slowly, and yet others with very high thresholds may never change at all. A classic threshold model, operationalized in the pre-computer era as a manual tabletop exercise, was Thomas Schelling's (1971, 1972) housing segregation model. Residents of a community are presumed to possess some threshold of in-group preference (homophilia) for living in a neighborhood amidst others who have the same characteristic, such as race, ethnicity, or income. If a person (ego) perceives that the number of neighbors with out-group characteristic(s) has become higher than ego's threshold, then the unhappy ego tries to remedy the situation by moving to an available dwelling in another neighborhood that has the requisite homogeneity. Schelling showed that even small levels of homophilia would lead people to move to housing areas occupied predominantly by similar others, eventually creating highly segregated residential patterns. In a knock-on or domino effect, some of ego's new neighbors may then discover that *their* thresholds have been breached; in turn, they flee to areas satisfying their in-group preferences. As collective segregation increases across the community, the number of dissatisfied agents declines until an equilibrium is achieved. However, Schelling's segregated world does not guarantee that everyone will live happily ever after among their new neighbors.

Mark Granovetter (1978) replaced Schelling's attribute thresholds with behavioral thresholds to explain how an ego's personal actions depends on how many other people are currently participating in that activity. Although his main example was the decision to join a riot, Granovetter indicated that threshold models should be applicable to other types of contagious and collective actions such as the adoption of innovations, the spreading of rumors and diseases, labor strikes, voting, going to college, and migration. Specifically, a threshold is "the number or proportion of others who must make one decision before a given actors does so" (p. 1420). If the observed value reaches or exceeds an actor's threshold, it adopts the activity thereby increasing the size of the group engaged in the collective action. If the personal threshold is not reached, the actor refrains from joining and the collective activity ceases to grow. Illustrating the choice of whether to participate in a riot, Granovetter argued that such binary decisions involve an actor's cost-benefit calculations that depend in part on how many others chose one of the two alternatives. "The cost to an individual of joining a riot declines as riot size increases, since the probability of being apprehended is smaller the larger the number involved" (p. 1422). An objective of the threshold model's mathematical equations was to compute, from the initial distribution of thresholds, the equilibrium outcome of the number or proportion of persons making the binary decisions. Another goal was to assess "the stability characteristics of any distribution's equilibrium under a variety of possible perturbations" (p. 1429). Granovetter's model did not try to explain how people acquire their thresholds, but takes a community's distribution of preferences as given and investigates the consequences following from that distribution. For some extensions and elaborations on threshold models, see Granovetter and Soong 1983, 1986, 1988; Chwe 1999; Grabisch and Li 2020; Wiedermann et al. 2020.

Granovetter modeled an individual's decision making using deterministic equations comparing the current level of rioting relative to the individual's threshold.

Michael Macy and Anna Evtushenko (2020) relaxed that assumption by introducing individual behavioral randomness. "Results showed that stochastic thresholds made little difference in small groups, but as group size increased, the idiosyncrasies of behavior at the micro level can have a surprisingly stabilizing effect on the unpredictability that arises from sampling errors at the macro level – the predictability paradox" (p. 643). Random errors in individual actions can trigger rapid behavioral cascades—people making choices solely on the decisions of others—which are less sensitive to group composition and more attuned to collective interests.

Threshold models of collective action through imitative behavior are a subset of behavioral contagion and innovation diffusion dynamics in social networks. Among the factors affecting diffusion, Thomas Valente and George Yon (2020) identified "characteristics of the initial adopters, the seeds; the structure of the network over which diffusion occurs; and the shape of the threshold distribution." They modeled the rate and prevalence of diffusion under seven seeding conditions, three network structures, and four threshold distributions with different variances. "Higher threshold distribution [standard deviations] were associated with faster and more prevalent diffusion. This association clearly indicates that variability in the threshold levels of the community/population has a profound effect on the speed and prevalence of diffusion" (p. 245). They urged researchers to conduct more experiments on threshold distributions for various behaviors and types of populations, which would enable public health officials to design more effective interventions for communities with many or few low- and high-threshold individuals.

Sometimes gatherings with the potential to turn violent avoid that fate. On June 1, 2020, many cities across the US saw rock-throwing, squad-car burning, arrests, injuries, and deaths as demonstrators expressed outrage over the killing of George Floyd by a Minneapolis policeman. In contrast, Newark, New Jersey, the scene of a deadly 1967 race riot, experienced a mostly peaceful 12,000-person march through the downtown. A key calming element was the 50-member Newark Community Street Team, formed 6 years earlier to try to de-escalate violence, along with other community groups "deployed throughout the crowd to try to isolate those intent on destruction" (Tully and Armstrong 2020). Those predominantly African-American youths sought to block violent actors from inciting imitators:

> At one point during the march, protesters lit an American flag on fire in the middle of Broad Street as a young man used a bat to strike a window of a Dunkin' Donuts store, witnesses said. "He hit the window one time and there was like 20 people standing in front of him," Mr. Sherrills said. As protesters whom he called "provocateurs" moved toward buildings owned by Prudential Financial, the city's most prominent business anchor, which has maintained a presence in Newark for 145 years, a similar standoff was defused, he said. (Tully and Armstrong 2020)

The deployment of civilian peacekeepers in efforts to defuse potential or actual violent situations is increasingly prevalent in both domestic protests and international armed conflicts (Whitehill et al. 2014; Julian and Schweitzer 2015; Bramsen 2019; Julian 2020; Morrell 2020). Applying principals of nonviolent resistance pioneered by Mahatma Gandhi and Martin Luther King, peacekeepers seek a de-escalation or cessation of violence by intervening between perpetrators and their targets. Peace teams are "embroiled in assemblages that include police, demonstrators and organizers, counter-protesters, and tools and technologies" (Canevez and Winter 2022). A threshold model of violence reduction may generate a "reverse bandwagon" effect where participants "opt-out" of their decisions to commit illegal acts when they observe others desisting. A survey of university students used hypothetical scenarios about intentions to fight and steal to find evidence of such opt-out thresholds (McGloin et al. 2021; see also McGloin and Rowan 2015; McEvoy 2010). The results indicate that some people may stop rioting after a number of others first decide to stop, while others may remain impervious to such a reverse bandwagon effect.

3 Rioting ABMs

Many agent-based models of collective violence simulate struggles between lawbreakers and law enforcement agents. A relative handful of ABMs concentrate on the microlevel decisions of agents to participate in or desist from rioting, the topic of this chapter.

When Kenya's 2007 election results were announced, rumors of rigged voting set off a 2-month cycle of sporadic violence in Kibera, an informal settlement in the capital city, Nairobi, leaving 1100 people dead and 350,000 internally displaced. Bianica Pires and Andrew Crooks (2017) created a complex ABM that integrated social network relations and geographic information systems (GIS) to show how a population and its physical environment interact with contagious rumors to trigger the ebb and flow of rioting. They ran a full-scale model, with 235,000 agents representing the estimated population of Kibera moving on a GIS grid approximating the locations of boundaries, facilities (businesses, schools, hospitals, religious institutions), and transportation routes of the settlement. The analysts systematically varied such parameters as commitment to others with the same social identity, traveling to school or work, the probability of losing employment, and individual decisions whether to commit violent acts or remain peaceful.

> [The model] demonstrates that the propagation of rumors through the unique, local interactions of agents via social networks can be simulated. By grounding the agents' cognitive framework in theory and applying empirical data to create a landscape that represents a real world location, micro-level interactions resulted in macro-level phenomena in the form of rioting. (Pires and Crooks 2017: 78)

Their results indicated that youths are more prone to rioting than older adults and that people frustrated from staying at home and being unemployed are more willing to aggress. "While this may not be a surprising result, it does demonstrate that the model is capturing the right types of dynamics, which is a form of validation" (p. 78). The aim of the ABM was not to replicate historical events accurately, but to capture the cyclical pattern of escalation and de-escalation of riots, leaving for later model modifications the exacerbating effects of police agents and rival groups.

Obtaining plausible parameter estimates by observing real illegal actions can be problematic. A team of psychologists creatively conducted experiments on college student subjects and used the results to construct an ABM (Dezecache et al. 2021).

A smartphone-based game, "Parklife," require physically co-present red and blue team members to tap screens to grow trees and flowerbeds in their team's virtual park. Players could also tap to vandalize the other team's park. The investigators secretly made teams unequal by requiring the disadvantaged players to tap more of each reward. "The experience of inequity caused the disadvantaged team to engage in more destruction, and to report higher relative deprivation and frustration" (p. 1). The subsequent ABM was designed to assess whether Parklife players vandalized at random, or because social comparison and frustration generated team norms.

> In the full model, our findings suggest that players focus mainly on their own team, and coordinate their behaviour by performing the opposite function of those on their team (i.e., if many team mates are working, they vandalize and vice versa). Participants balance team behaviour between working and vandalizing, providing evidence of coordinated behaviour across the teams. (Dezecache et al. 2021: 5)

Although frustration was the main variable that increased vandalism by the disadvantaged group, social identification and emergent group norms were also important factors. The interaction between individual-level frustration and team behavior (working or vandalizing) better captured the experimental data than frustration alone.

Alastair Clements and Nabil Fadai (2022) developed a stochastic ABM of a sports riot with two sub-populations of agents—rioters and bystanders—and individual-level mechanisms for recruitment and defection to the other subpopulation. Model dynamics vary with location population density, the number of rioters or bystanders occupying the four von Neumann neighborhoods (North, South, East, West) around an agent's patch. (A patch cannot be occupied by more than one agent at a time.) For example, if a bystander encounters at least two rioters in its neighborhood, the bystander gets recruited to rioting. The analysts ran multiple simulations under three initial conditions: (1) Mild Unrest: rioters can become bystanders regardless of the number of bystanders nearby, but bystanders are recruited only when two or more rioters are present. (2) Moderate Unrest: rioters and bystanders both defect in the presence of one agent of the opposite sub-population. (3) Severe Unrest: bystanders become rioters regardless of how many rioters are nearby, while rioters only defect when two or more bystanders are nearby. Thus, in each scenario, the recruitment-defection thresholds are identical for all members of the two sub-populations.

> [These] initial conditions give rise to three main qualitative features: complete take-over by rioters, complete take-over of bystanders, or a co-existence equilibrium of both sub-populations. All three qualitative features are faithfully reproduced in the continuum limit of the ABM, which also gives rise to a systematic method of relating individual-level recruitment and defection rates to their analogous population-level counterparts. (Clements and Fadai 2022: 10)

To ensure computational efficiency, the model's algorithm allowed the researchers to track only the aggregate numbers of rioters and bystanders, rather than tracking agents individually. Further model extensions could count the number of times each agent switched from one sub-population to the other and back again, thus obtaining more precise estimates of the underlying recruitment and defection rates that give rise to the global rates.

4 What a Riot! ABM

"What a Riot!" is a NetLogo ABM of a crowd of bystanders moving around a public area such as a park, street corner, or town square. At the start, the program randomly assigns an integer to every bystander that is a threshold value for joining or defecting from riotous behavior. A bystander is recruited to become a rioter if the number of nearby instigators equals or exceeds its threshold. A recruited rioter will defect back to bystander status if the number of nearby peacekeepers equals or exceeds its threshold. The model puts no limit on the number of times a bystander may be recruited to or dissuaded from rioting. The names of the four types of agents, their colors, and behaviors are:

(1) **Bystanders** (green) are nonviolent agents at a public site. Every bystander has a randomly assigned threshold integer for rioting or defecting, which is the number of rioters or peacekeepers around the bystander's current location. The N-Threshold slider allows the investigator to decide how many integers, between 1 and 10, will be randomly assigned. For example, setting the threshold slider at 3 mean that approximately one-third of the bystanders have thresholds of 1, 2, or 3.

(2) **Instigators** (red) are initiators of riot actions whose thresholds are 0. That is, they do not need to see anyone else rioting before engaging in physical violence against property or people, such as breaking store windows, fire-bombing buildings, or hitting people with sticks or stones. (The specific types of riot activities are not specified in the ABM.)

(3) **Rioters** (orange) are initially bystanders that turned violent when they encountered a number of instigators, within a 1-patch radius around their current location, that equals or exceeds their threshold value. When a bystander joins the riot, its color changes from green to orange to indicate it's engaging in riotous behavior. The presence of rioters near bystanders does contribute to recruitment; only nearby instigators count toward a bystander's threshold for becoming a rioter.

(4) **Peacemakers** (blue) seek to persuade rioters to desist through a reverse imitation process. If the number of peacemakers in a 1-patch radius around a rioter equals or exceeds its threshold, the rioter ceases to imitate the instigators and its color changes from orange to green. However, a former rioter is susceptible to reconversion if the number of neighboring instigators again meets or exceeds the threshold.

Sliders on the interface allow a researcher to decide the numbers of agents, change the number of thresholds (higher threshold values could be rarely or even impossibly achieved), and set the percentage of bystanders-turned-rioters that will terminate a model run.

Two screen captures illustrate simulation outcomes for different parameter configurations. In Fig. 1, the 300 bystanders, whose thresholds were randomly assigned to be 1 or 2 nearby instigators (approximately half the bystanders have one or the other of the two thresholds). This experiment has 10 instigators randomly moving around that convert bystander to rioter when their numbers reach or exceed a nearby bystander's threshold. With no peacemakers to reverse recruitment in this run, the

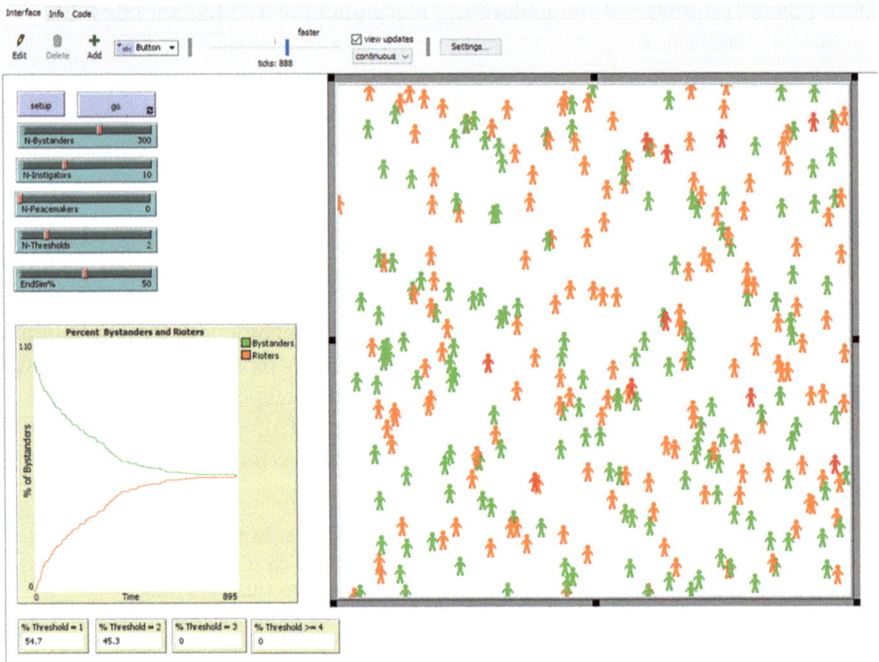

Fig. 1 What a Riot! ABM with two bystander thresholds

simulation halted, at 1036 ticks, when half the bystanders had become rioters. In Fig. 2, the only parameter change was to assign the bystanders randomly to three thresholds levels for recruitment by nearby instigators (1, 2, or 3, approximately one-third of the bystanders in each level). The duration of this model run was ten times longer, terminating after 10,779 ticks, presumably because seeing three nearby instigators was a more uncommon experience.

Table 1 summarizes the time, in ticks, when half of the bystanders became rioters for 200 repetitions with the model parameters set in the two examples in Figs. 1 and 2. Raising the thresholds for one third of the bystanders increased more than 13-fold the median and mean times when half the bystanders became rioters. These experiments demonstrate the relative infrequency of multiple instigators simultaneously moving into close proximity to bystanders who have higher recruitment thresholds.

The third run, in Fig. 3, set the bystander thresholds at three, and added 10 peacemakers to counteract the incitement of 10 instigators. After an initial increase in rioters, the percentage of rioters stabilized, fluctuated without apparent trend, and

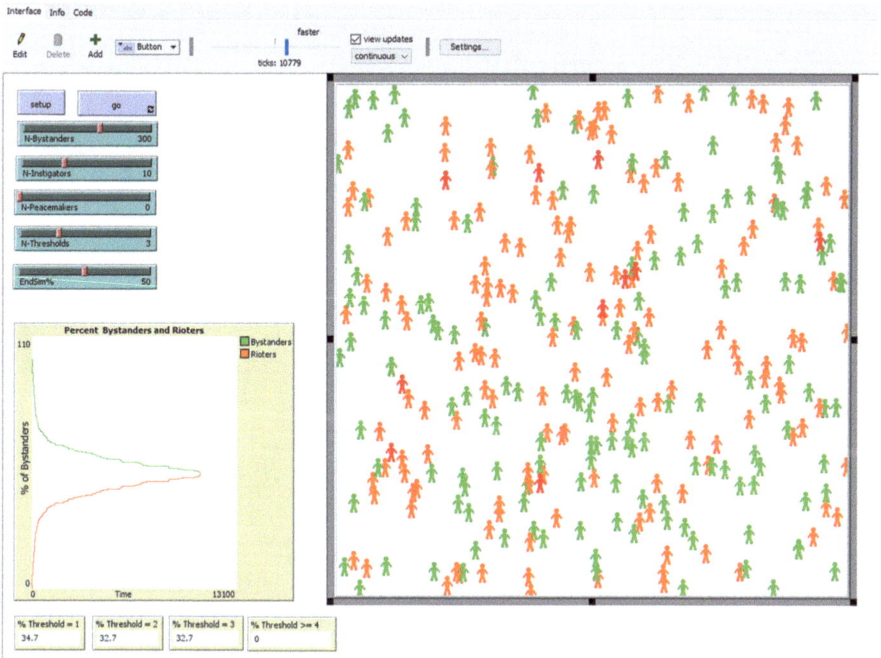

Fig. 2 What a Riot! ABM with three bystander thresholds

Table 1 Two experiments with different thresholds

	Model 5.1	Model 5.2
Number of thresholds	2	3
Minimum ticks	557	6662
Maximum ticks	4341	26,215
Median ticks	1071	14,449
Mean ticks	1211	14,619
Standard deviation	507	3703
N of repetitions	200	200

never reached 50% of the population even after more than 100,000 ticks. In a 200-repetition experiment, all failed to turn 50% of bystanders into rioters before 50,000 ticks, at which time the run was cancelled. In configurations where a substantial proportion of the bystanders have high recruitment thresholds, a sizeable contingent of peacemakers who promote rioter defections back into bystanders appears sufficient to prevent more widespread rioting.

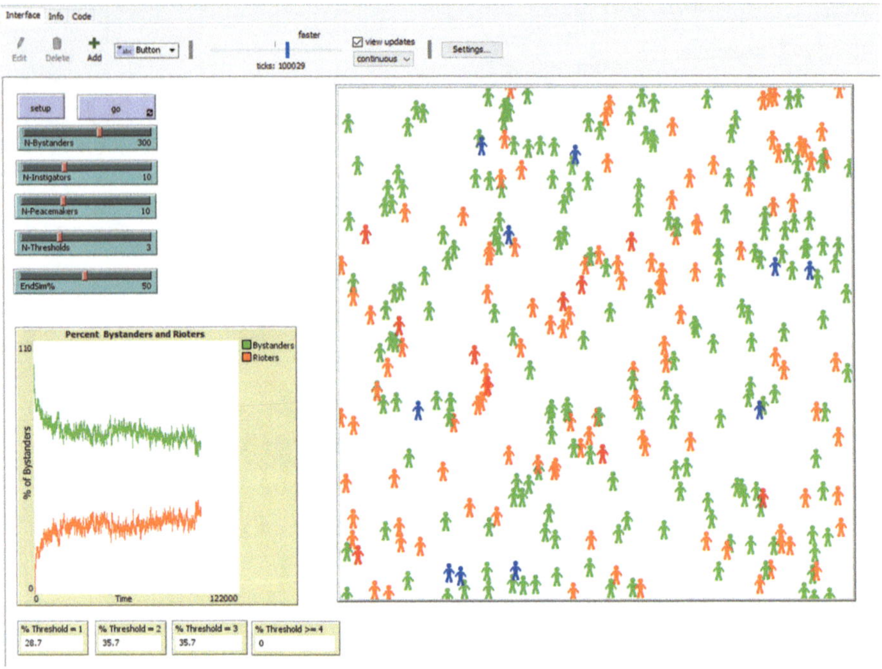

Fig. 3 What a Riot! ABM with three bystander thresholds and peacekeepers

5 Discussion

At its core, the spontaneous diffusion of riotous behavior is a "monkey see, monkey do" imitation process. Sports fans will riot when their team loses a big game, when their team wins, and even before the game's outcome is known. What are the causes—frustration, overexcitement, jubilation, alcohol, adrenaline, testosterone, the madness of crowds? The underlying reasons are plausibly as diverse as the numbers of rioting individuals. Attempting to incorporate explicit causal mechanisms that capture every individual participant's psychological motives is futile. Fortunately, agent-based models of riots do not require complex details. Simple behavioral decisions about when to join or leave a riot are analogous to the ABMs discussed in chapter "Infecting", where susceptible people come into direct contact with an infectious person and become infected themselves. The main difference in the transmission processes of the two ABMs is that, instead of a virus transmitted between people with some probability, bystanders at a riot imitate illegal actions when exposed to preceding miscreants. Pandemics and riots accelerate rapidly through a community or a crowd as the newly infected and the recently recruited pass their disorders on to others. Similarly, disease and misconduct can be brought under control by interventions designed to avoid exposure or to reverse contamination: masking and social distance during a pandemic, counter-messaging by peacekeepers during a riot.

The motives and goals of riot instigators and peacemakers are also not specified in the ABM. Some troublemakers might be thrill-seeking pyromaniacs who get turned on by flames. Others may get swept up by anger or elation in the heat of moment, acting out fantasies they later will come to regret inside a jail cell. Still others could be *agents provocateur* sent by law enforcement to infiltrate an organization and incite its members to commit illegal actions that would result in their arrest and incarceration. For example, the Federal Bureau of Investigation used an informant who, gaining prominence in the Denver Black Lives Matter movement, allegedly "encouraged protesters to engage in increasingly violent demonstrations while trying to entrap them in criminal misdeeds" (Pilkington 2023). On the other hand, peacemakers seek to enact a rational agenda based on training and experience to defuse rowdiness before it gets out of hand.

This chapter examined collective actions that did not include reactions by law enforcement officials after a riot breaks out. Sometimes police or armed forces succeed in quickly tamping down violence, but at other times their aggressive missteps spark larger conflagrations that consumes more property and lives. Chapters "Insurrecting" and "Insurging" examine ABMs involving violent conflicts between opposing groups.

References

Adams JJ (2021) Stanley cup riot, 10 years later; 'how come we were able to make the right choice and everybody else was not?'. Vancouver Sun, 12 June. https://vancouversun.com/sports/hockey/nhl/vancouver-canucks/how-come-we-were-able-to-make-the-right-choice-and-everybody-else-was-not. Accessed 1 June 2022

Andres L, Fabel M, Rainer H (2023) How much violence does football hooliganism cause? J Public Econ 225:1–15

Banel F (2021) Release the Canucks: remembering the Stanley cup riot of June 2011. Accessed 31 May 2022

Bramsen I (2019) Avoiding violence: eleven ways activists can confine violence in civil resistance campaigns. Confl Resolut Q 36:329–344

Canevez R, Winter J (2022) Peace teams in the protest-repression nexus: a sociomaterial perspective of de-escalatory tactics. In: Paper presented at the 55th Annual Hawaii international conference on system sciences. University of Hawai'i at Mānoa, Honolulu. https://scholarspace.manoa.hawaii.edu/server/api/core/bitstreams/6bd86124-7c75-4789-8bd1-d104e821cb38/content. Accessed 31 May 2022

Chwe MS-Y (1999) Structure and strategy in collective action. Am J Sociol 105:128–156

Clements AJ, Fadai NT (2022) Agent-based modelling of sports riots. Physica A Stat Mech Appl 597:127279

Dezecache G, Allen JM, von Zimmermann J, Richardson DC (2021) We predict a riot: inequity, relative deprivation and collective destruction in the laboratory. Proc R Soc B 288(1959):20203091

Dunning E (2000) Towards a sociological understanding of football hooliganism as a world phenomenon. Eur J Crim Policy Res 8:141–162

Grabisch M, Li F (2020) Anti-conformism in the threshold model of collective behavior. Dyn Games Appl 10:444–477

Granovetter M (1978) Threshold models of collective behavior. Am J Sociol 83:1420–1443

Granovetter M, Soong R (1983) Threshold models of diffusion and collective behavior. J Math Sociol 9:165–179
Granovetter M, Soong R (1986) Threshold models of interpersonal effects in consumer demand. J Econ Behav Organ 7:83–99
Granovetter M, Soong R (1988) Threshold models of diversity: Chinese restaurants, residential segregation, and the spiral of silence. Sociol Methodol 18:69–104
Holdo M, Bengtsson B (2020) Marginalization and riots: a rationalistic explanation of urban unrest. Hous Theory Soc 37:162–179
Julian R (2020) The transformative impact of unarmed civilian peacekeeping. Glob Soc 34:99–111
Julian R, Schweitzer C (2015) The origins and development of unarmed civilian peacekeeping. Peace Rev 27:1–8
Macy MW, Evtushenko A (2020) Threshold models of collective behavior II: the predictability paradox and spontaneous instigation. Sociol Sci 7:628–648
McEvoy DM (2010) Not it: opting out of voluntary coalitions that provide a public good. Public Choice 142:9–23
McGloin JM, Rowan ZR (2015) Threshold model of collective crime. Criminology 53:484–512
McGloin JM, Thomas KJ, Rowan ZR, Deitzer JR (2021) Can the group disincentivize offending? Considering opt-out thresholds and decision reversals. Criminology 59:738–765
Morrell L (2020) From Iraq to Minneapolis, nonviolence group worked to prevent voter intimidation during election. Minnesota Daily, 14 November. https://mndaily.com/263851/news/from-iraq-to-minneapolis-nonviolence-group-worked-to-prevent-voter-intimidation-during-election. Accessed 6 June 2023
Ostrowsky MK (2018) Sport fans, alcohol use, and violent behavior: a sociological review. Trauma Violence Abuse 19:406–419
Pilkington E (2023). Fears of renewed FBI abuse of power after informant infiltrated BLM protests. The Guardian, 14 February. www.theguardian.com/us-news/2023/feb/14/fbi-abuse-of-power-alleged-informant-denver-blm-protests. Accessed 10 June 2023
Pires B, Crooks AT (2017) Modeling the emergence of riots: a geosimulation approach. Comput Environ Urban Syst 61:66–80
Schelling TC (1971) Dynamic models of segregation. J Math Sociol 1:143–186
Schelling TC (1972) A process of residential segregation: neighborhood tipping. In: Pascal AH (ed) Racial discrimination in economic life. D.C. Heath, Lexington, pp 157–114
Singh D (2021) Where society fails: inside the Vancouver Stanley cup riot. Sportsnet.ca. www.sportsnet.ca/nhl/longform/society-fails-inside-vancouver-stanley-cup-riot. Accessed 31 May 2022
Tully T, Armstrong K (2020) How a city once consumed by civil unrest has kept protests peaceful. New York Times, 1 June. www.nytimes.com/2020/06/01/nyregion/newark-peaceful-protests-george-floyd.html. Accessed 25 May 2022
Valente TW, Vega Yon GG (2020) Diffusion/contagion processes on social networks. Health Educ Behav 47:235–248
Whitehill JM, Webster DW, Frattaroli S, Parker EM (2014) Interrupting violence: how the CeaseFire program prevents imminent gun violence through conflict mediation. J Urban Health 91:84–95
Wiedermann M, Keith Smith E, Heitzig J, Donges JF (2020) A network-based microfoundation of Granovetter's threshold model for social tipping. Sci Rep 10:1–10
Zimniak R (2020) The sociological and psychological aspect of football hooliganism. Teisė 117:138–151

Recruiting

TiffanyWu

Keywords Alter · Deprivation theory · Ego centric network · Grievance theory · Recruiting Movement Supporters ABM · Social movement · Social movement organization · Social network · Strength of tie

1 Introduction

After George Floyd was murdered in 2020 by a Minneapolis police officer kneeling on his neck during an arrest, many rallies, vigils, and marches in support of the Black Lives Matter movement sprang up in at least 140 U.S. cities and all 50 states (Taylor 2021). Most protests were peaceful, but by June about 10,000 people had been arrested on charges ranging from curfew violations to burglary and looting (Snow 2020). Polling data estimated between 15 and 26 million Americans participated in demonstrations, about half of them for the first time (Buchanan et al. 2020). Unlike previous Black Lives Matter protests starting in 2013, a majority of the George Floyd protesters were white, young, and affluent:

> "Without gainsaying the reality and significance of generalized white support for the movement in the early 1960s, the number of whites who were active in a sustained way in the struggle were comparatively few, and certainly nothing like the percentages we have seen taking part in recent weeks," said Douglas McAdam, an emeritus professor at Stanford University who studies social movements. (Buchanan et al. 2020)

The original version of the chapter has been revised. A correction to this chapter can be found at https://doi.org/10.1007/978-3-031-86199-4_10

Supplementary Information The online version contains supplementary material available at https://doi.org/10.1007/978-3-031-86199-4_6.

D. Knoke, *Network Collective Action*, Lecture Notes in Social Networks,
https://doi.org/10.1007/978-3-031-86199-4_6

Despite the global pandemic, protests spread to dozens of cities and 60 countries around the world. Huge crowds gathered in solidarity in Auckland, Berlin, Copenhagen, Dakar, Istanbul, London, Melbourne, Montreal, Nairobi, Rio de Janeiro, Seoul, and Tokyo (McCurry et al. 2020; Taylor 2020). World leaders denounced racism and injustice, with United Nations Secretary-General António Guterres tweeting, "Racism is an abhorrence that we must all reject. Leaders in all sectors of society must invest in social cohesion so every group feels valued. That means addressing inequality and discrimination, strengthening support for the most vulnerable and providing opportunities for everyone" (United Nations' Post 2020). The eventual conviction and imprisonment of the four cops involved in Floyd's murder gave some hope that justice prevailed in his case—even as fatal police shootings of unarmed minoritized American citizens continued apace.

This chapter examines the initial step in a person's recruitment to a social movement: moving from a neutral bystander or passive sympathizer to becoming a committed member. Subsequent steps involve participation in movement activities, such as donating money and time, attending meetings, and engaging in public demonstrations. These later phases are the subjective of chapters "Insurrecting" and "Insurging", where protesters confront opposing forces that may result in arrests, beatings, and death. After reviewing the theoretical and empirical research literature on joining social movements, we present an agent-based model of movement recruitment through network connections.

2 Social Movements

Social movements are collective efforts by disadvantaged and relatively powerless people who use extra-institutional means to promote or resist social changes in some societal domain, such as politics, economics, culture, race and ethnicity, or sexual identity. Movement activists deploy a diverse repertoire of tactics, some emphasizing confrontation with authorities and disruption of conventional and routine activities. Many tactics involve nonviolent ways to publicize the movement's demands, including collecting signatures on petitions, holding press conferences, posting placards, managing Websites, conducting teach-ins, sit-ins, and hunger strikes, holding rallies and protest marches. However, extremist members of a movement may sometimes resort to violent actions—breaking windows, torching buildings and automobiles, assaults, street fights, riots, bank robberies, bombings, assassinations. Law enforcement officers or counter-protesters may foist violence on nonviolent movement members; for example, "pro-life" radicals who murder abortion clinic doctors.

A social movement organization (SMO) is a named formal organization engaged in actions to advance a movement's goals. Movements often have multiple SMOs with diverse and overlapping goals and collective actions. For example, the environmental movement comprises such organizations as Audubon Society, Greenpeace, Earth Now!, Friends of Earth, Nature Conservancy, Natural Resources Defense Council, Sierra Club, and World Wildlife Federation. It even includes the secretive

Earth Liberation Front (ELF), designated as an eco-terrorist organization by the Federal Bureau of Investigation (Brown 2020).

Resource mobilization theories of social movement formation and collective action emphasize that a movement's success in attaining its goals depends substantially on obtaining valuable resources (Klandermans 1986; Edwards et al. 2018; Angelopoulos et al. 2023). Resources—money, bodies, leadership, skills, publicity, recognition, legitimacy—can be provided by activists, casual participants, passive donors, and sympathetic organizations, such as churches, foundations, voluntary associations, labor unions, social media, colleges, even political parties. The greater the total amount of resources available to a social movement, the more likely are new SMOs to emerge that simultaneously collaborate and compete for those resources (McCarthy and Zald 1977, 2001; Zald and McCarthy 2002). Resource mobilization by SMOs bears striking parallels to business entrepreneurs trying to raise investment capital and open new market niches.

Explanations of why people join social movements and participate in collective action events range from deprivation and grievances to aspiration and altruism (Knoke and Wisely 1990). Earlier theories emphasized individual psychological dispositions: people who experience grievance, stress, or frustration would somehow come together and agitate for public authorities to do something to alleviate their maddening circumstances (LeBon 1896; Turner and Killian 1972; Opp 1988). Subsequent empirical evidence indicated that while dispositional propensities may be a necessary prerequisite, they are insufficient to transform grievances into action. Movements can try to persuade strangers to join by impersonal means, such as mass communications, door-to-door canvassing, phone banks, and teach-ins. However, these methods are often inefficient compared to personal relations with movement activists.

A more effective way to expand a movement's ranks is to tap into current members' preexisting network ties to others, such as family, friends, coworkers, and voluntary associations. Strong emotional connections to others (alters) in egocentric networks increase the probability of recruitment to social movements (Tindall et al. 2021). Movement activists target others whose shared social identities and sympathy toward a movement's values and goals could persuade them to join and provide resources to SMOs. An analysis of political and religious movements found that "the probability of being recruited into a particular movement is largely a function of two conditions: (1) links to one or more movement members through a preexisting or emergent interpersonal tie; and (2) the absence of countervailing networks" (Snow et al. 1980: 798). Recruitment is a function of differential availability, that is, "how tightly individuals are tied to alternative networks and thus have commitments that hinder the recruitment efforts of social movement organizations" (p. 794). People working in high-demand jobs or raising several children are less available than students, unmarried young adults, or retirees to accept an "invitation" from a network contact to attend a movement event.

Strong network ties are particularly crucial in high-risk and high-cost situations that are likely to deter weakly attached sympathizers. When the perceived chances of success are low and the potential for harm is high (e.g., from police violence or losing one's job), a rational decision is not to get involved. Network relations can overcome such negative rational calculations, especially if people value keeping or strengthening their relations with a movement's activists in their personal networks. In a series of articles on the importance of network recruitment, Doug McAdam analyzed applications to the Student Nonviolent Coordinating Committee's 1964 Mississippi Freedom Summer, a campaign to register Black voters in the segregated South (McAdam 1986, 1992; Fernandez and McAdam 1988; McAdam and Paulsen 1993). The event was exceptionally high-risk—three civil rights activists were abducted and murdered by the Ku Klux Klan that summer. McAdam conducted a content analysis of responses by 961 people to an application form question about why they "would like to work in Mississippi." He coded information about motives, beliefs, ideologies, and connections to organizations and individual activists in the civil rights movement. Compared to 241 applicants who withdrew, the 720 men and women who travelled to Mississippi had more organizational affiliations, higher levels of prior civil rights activity, more extensive and stronger current or past ties to civil rights activists. The applicants were also asked to list at least 10 persons whom they wished to be kept informed about their summer activities:

> The interesting finding is that participants supplied many more names of other participants and known activists than did withdrawals. The differences are especially pronounced in the two strong tie categories, with participants listing more than twice the number of volunteers and nearly three times the number of activists as the withdrawals. (McAdam 1986: 79-80)

In contrast to the diffusion of information via weak ties (Granovetter 1973), McAdam argued that strong ties are more likely to influence others to join in high-risk activities. "Having a close friend engage in some behavior is likely to have more of an effect on someone than if a friend of a friend engages in that same behavior" (1986: 80). Stronger social bonds help social movement recruits to persevere in collective actions undertaken in a hostile environment. Subsequent research underscoring the importance of social networks in movement recruitment include studies of the Arab Spring uprisings in 2010 (González-Bailón et al. 2011); Islamic State radicalization (Chatfield et al. 2015); a climate change demonstration (Van Laer 2017); Black Lives Matter and Women's Rights SMOs (Bozarth and Budak 2017); protest in Paris after murders at the satirical magazine Charlie Hebdo (Larson et al. 2019); the environmental movement (Saunders 2022); and a protest parade against governmental austerity policies (Walgrave and Wouters 2022). The recent rapid rise of social media-based Internet activism provided traditional social movement organizations with additional recruitment equipment for their toolkits.

3 Social Movement Simulations

Although many researchers have proposed agent-based models of social movement collective actions, only a handful have examined the recruitment process. Those efforts include models of: recruitment by "activators" that result in a collective good (Mosler and Tobias 2001); group formation by agents inviting network neighbors to join (Geard and Bullock 2008, 2010); stronger network ties exerting greater influence on potential recruits (Hu et al. 2015); worker strikes undermined by distrust among workers of differing ethnicities and cultures (Kim and Hanneman 2011); and student protests increased with the degree distribution of friendship network ties (Raphiri et al. 2023). Space constraints allow overviews of three exemplary ABMs.

Navid Hassanpour (2010) argued that the structure of social networks and updating beliefs on thresholds are twin factors affecting mobilizations for social movement collective action. He criticized Mark Granovetter's threshold model for its unrealistic assumption that levels of participation are fully visible to all others. Instead, he argued that individual thresholds "are strongly influenced by the network of local connections and spatial confines of the situation in which the individual is situated" (p. 5). Although Hassanpour did not write a formal agent-based model of movement mobilization, his equations predicting an ebb-and-flow oscillation of protest participation levels could readily be programmed as an ABM that combines local network configurations and agents' dispositions to protest or desist contingent on perceived proclivity of network alters' thresholds. In subsequent research on the 2011 Egyptian protests, Hassanpour (2016) found that the escalation of protest originated in a peripheral network rather than through centralized activists after a communication blackout was imposed by the Mubarak regime. A similar small world network dynamic arose when the 2011 widespread Syrian pro-democracy protests against the Assad regime led to a prolonged civil war. The book's title, *Leading from the Periphery and Network Collective Action*, neatly encapsulates his core hypothesis that a peripherally positioned network of risk-taking agents are better able that a central network to launch a successful cascade of revolutionary collective action.

A team at the Rand Corporation explored the role of information and communication technologies (ICT) in recruitment based on ABMs of three social movements: the Arab Spring uprisings in Egypt and Syria and the 2019 Hong Kong protests against a government bill to allow extraditions to mainland China (Frank et al. 2022; Marcinek et al. 2022). For the Egyptian case, the team investigated how "social media helped the Egyptian protesters acquire international support for the protests, recruit new protesters into the movement, organize specific protest events, and communicate live updates to the protesters" (Frank et al. 2022: 12). Their ABM has three types of agents: authorities and citizens who are either protesters or non-protesters. At the beginning of the simulation, no connections exist among the citizens. As the process unfolds, citizens randomly form links to one another through two ICT mechanisms, "outreach" and "networking." Outreach links occur if ego and an alter are both protesters or they have similar attributes (identity homophily). The networking mechanism creates a link between ego and the alter of an alter when both are protesters or have similar identities. The investigator can manipulate outreach and networking parameters to simulate raising or lowering the costs of communication. An authority agent can surveil both types of citizens through its control over the ICTs used for citizen communication. The analysts ran their model numerous times with varied configurations of protest dynamics and surveillance. Both ICT mechanisms had positive effects on the speed with which protesters connected to one another. The effects depended on the initial size of the protesting population. For a small number of protesters, recruitment through outreach was quicker, but in a larger population, recruitment through networking led to faster link formation. The researchers drew two broad conclusions;

> First, the model reflects a common observation of sociologists and political scientists who try to predict how much support a given social movement might receive. Namely, depending on the underlying characteristics of the population, the final number of participants can significantly vary. Secondly, the model indicates that in the initial phase of social movement formation, technologies that enable casting a wide net accelerate both the social movement growth and the network formation more than those allowing elite, invitation-only membership. (Marcinek et al. 2022: 21)

A challenge for future extensions of the ICT is to increase its realism by incorporating both authority and participant objectives and actions into the ABM.

The Social Identity Model of Collective Action (SIMCA), an ABM implemented in Python by Alexander Petrov, Andrei Akhremenko, and Sergey Zheglov (2023), examined the effects of dual identities on protest turnout against autocratic regimes. Opposition members have both a narrow identity with a specific dissent group and a broad identity with a second opposition group in an anti-regime coalition. The dynamics of these two identity components depend on whether the regime's strategy aims to repress one or two groups. Each individual decides whether to attend today's rally depending the previous day's rally turnout, the regime's repression severity and its subjugation strategy (targeted or broad), and attendance by her network friends, taking into account their partisanship composition. Other micro-level variables include anger and risk aversion. Among the major findings of the simulations: regime repression can increase turnout uncertainty; targeted repression is unable to suppress the protests; and network homophily affects turnout of the surviving campaigns "more than it affects the chance of surviving in the case of targeted repression and vice versa in the case of broad repression" (p. 2270). A limitation of the model is the short-term nature of its dynamics. Although government repression can decrease protest turnout, it tends to result in a longer-term rise of protest movements, civil violence, civil war, or coup d' état.

4 Recruiting Movement Supporters ABM

Our ABM takes the perspective of activist members of a social movement who attempt to recruit other persons in a network to support the movement, for example by donating money or participating in movement events. We don't specify details of the mechanisms used by the recruiter, such as the content of communications with targets, in persuading them to join. Success or failure in recruiting depends on which of three dispositions an agent has:

(1) **Instigation** – if agents have favorable views of the social movement's goals and actions, they are supporters who attempt to persuade others in their network to become supporters

(2) **Resistance** – if agents have negative or indifferent views toward movement, they cannot be converted into supporters and obviously won't try to recruit other network members to become supporters.

(3) **Neutrality** – if an agent is not hostile toward the social movement, it is potentially available for recruitment by movement activists. However, if it has no direct ties to any instigators, it is unavailable to be recruited.

A network's structural configuration affects the extent to which its members can be recruited. For example, a low-density network with few resisters may block activists from reaching and converting many neutral members. In contrast, a high-density network can result in a recruitment cascade because activists have multiple paths around any resister to reach and convert neutral agents into movement supporters.

Figure 1 shows the graphic display at the start of an experiment, after clicking the "Setup" button. The network was randomly generated from instructions to agents to create an average number of undirect ties (degrees) with other nearby agents. This experiment was set at 30 agents (people), with an mean of four undirected ties, and a single instigator of recruitment (colored green) located near the bottom. Another slider randomly assigns every agent a resistance number between 0 and 100. In this example, the threshold number was set at 20, which means that an agent whose random number is 20 or lower cannot be recruited whenever the resister becomes directly connected to a green movement supporter. The eight resisters in Fig. 1 are colored red. The 21 other agents, who are available to be converted into supporters, are colored blue.

Action begins by repeatedly clicking a "Go Once" button to observe the results of each "tick" of the time clock. A blue agent will be recruited and turn green if it has a direct link to an agent that has turned green in the preceding time interval. However, if a blue agent never comes into contact with a green agent, because all paths are blocked by red resisters, it remains blue. The experiment will end with

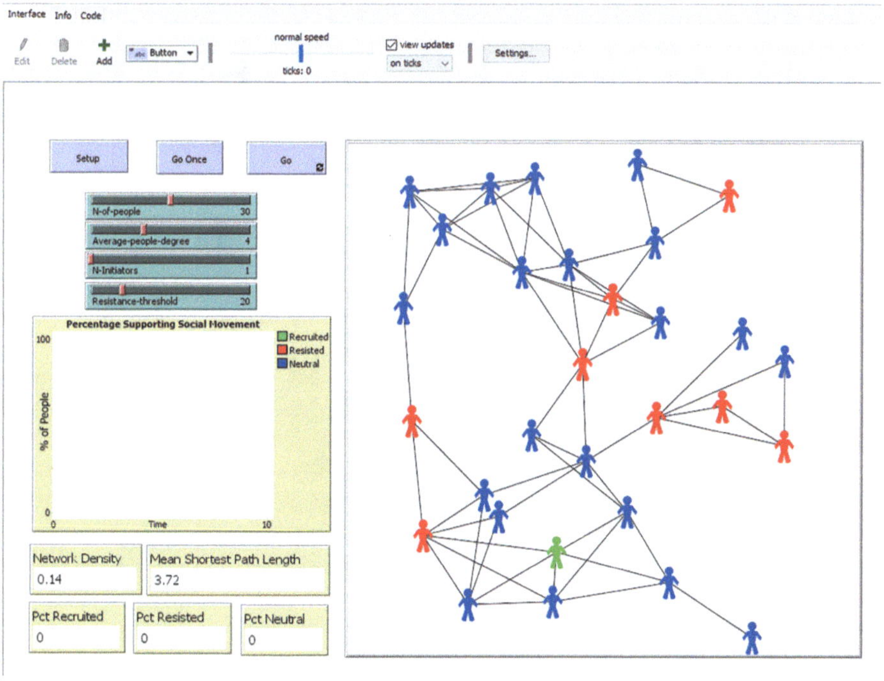

Fig. 1 Recruiting Movement Supporters ABM at start

some mixture of blue, green, and red agents. In the upper half of Fig. 1, notice that several red resisters appear to form a firewall that will prevent green supporters from directly contacting neutral blue agents. Can you anticipate which blue agents will and won't turn green? Figure 2 shows the outcome at the end of the experiment, after only two ticks of the clock. Nine blue agents on the bottom of the diagram were recruited to support the movement (all are now colored green), but the 12 neutral agents on the top were not (they remain blue). The plot on the left shows the changing network composition over the experiment's brief run. Experiments can be run multiple times with the same parameter settings to observe how common or rare are particular outcomes such as Fig. 2.

Table 1 summarizes the outcomes of running 200 experiments with two parameterizations of Recruiting Movement Supporters. Both models consist of 30 agents with the resistance threshold set at 20. In Model 6.1 the networks density (average degree) was set to 3, in Model 6.2, the density was raised to 5. As anticipated, given a fixed network size and resistance threshold, the lower density network recruited substantially fewer supporters (46.9%) than the higher density network (71.3%). The simple interpretation is that more connections give activists more paths to invite neutral members without access becoming blocked by resisters.

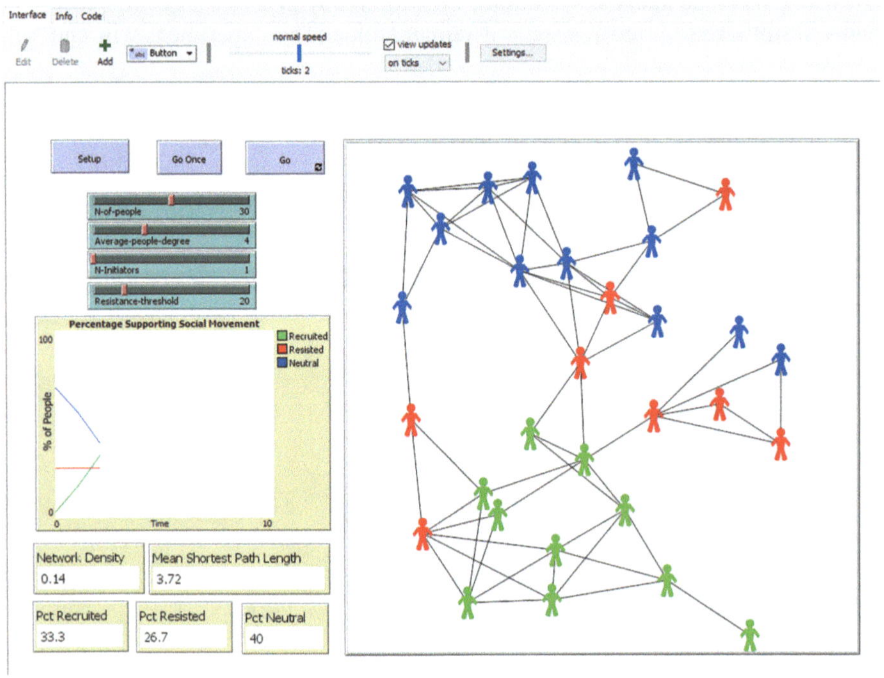

Fig. 2 Recruiting Movement Supporters ABM at end

Table 1 Two Recruiting Movement Supporters experiments with different network densities

	Model 6.1	Model 6.2
Network size (number of agents)	30	30
Average degree	3	5
Resistance threshold	20	20
Mean percentage recruited	46.9%	71.3%
Mean percentage resisted	19.9%	20.3%
Mean percentage neutral	30.7%	7.9%
Mean ticks	5.2	5.2
N of experimental runs	200	200

5 Discussion

The key insight from Recruiting Movement Supporters is that social structures matter. Microlevel networks of personal relations affect the macrolevel success or failure of a mobilization campaign. A low-density network that contains many members who are opposed or uninterested in supporting a social movement will result in recruitment fizzling out. In contrast, a high-density network provides alternate paths to bypass resisters, propagating a swelling recruitment wave across the network. Future experiments can assess the effects of differing network sizes and varied resistance thresholds. Other parameters to manipulate include multiple instigators and enabling newly recruited supporters to relapse into neutral status. A major change in the model would allow resisters also to recruit neutrals to the resistance. Unionization campaigns to certify or decertify a union as a workplace collective bargaining representative frequently pit pro- and anti-union employees in bitter confrontations. Historical instances of social movement factions, rivals, and schisms between moderates and radicals are the Bolsheviks and Mensheviks in the Russian Revolution, Irish Republican Army and Provisional IRA, and Fatah and Hamas in Palestine.

An elaboration of simple social movement recruitment ABM might distinguish among the types of incentives offered to potential supporters. As discussed in chapter "Theorizing", Mancur Olson (1965) argued that voluntary organizations providing only public goods generate suboptimal amounts of resources because rational actors will take a free ride. The solution is also to offer a variety of selective incentives available contingent on contributing time and money for collective action (Knoke and Wright-Isak 1982; Knoke 1988). Examples are magazine subscriptions, insurance, workshops, conferences, social gatherings, recognition ceremonies, and travel discounts. The challenge for ABM research is to designate agents with heterogenous preferences for various public goods and selective incentives, then measure their contributions to the social movement. If Olson and others are correct, more resources will be forthcoming under generous organizational provision of selective goods and services.

A final observation about social movement recruitment: it's just one of many network-based ABMs; for example, rumors and gossip (Alassad et al. 2023), word-of-mouth marketing (Hu et al. 2019), diffusion of innovations (Summad et al. 2023), supply chain formation (Colon et al. 2021). Although their substantive contents varies widely, these applications share an assumption that both micro- and macro-level behaviors must explicitly model network transmission dynamics. By investigating social relations among agents, analysts can improve knowledge about collective action and outcomes.

References

Alassad M, Hussain MN, Agarwal N (2023) Developing an agent-based model to minimize spreading of malicious information in dynamic social networks. Comput Math Organ Theory 29:487–502

Angelopoulos S, Canhilal KS, Hawkins MA (2023) From groups to communities: a resource mobilization theory perspective on the emergence of communities. Inf Syst Front:1–18

Bozarth L, Budak C (2017) Social movement organizations in online movements. Soc Sci Res Netw 3068546

Brown JM (2020) Notes to the underground: credit claiming and organizing in the earth liberation front. Terror Polit Violenc 32:237–256

Buchanan L, Bui Q, Patel JK (2020) Black lives matter may be the largest movement in U.S. history. New York Times, 3 July

Chatfield AT, Reddick CG, Brajawidagda U (2015) Tweeting propaganda, radicalization and recruitment: Islamic state supporters multi-sided twitter networks. In: Proceedings of the 16th annual international conference on digital government research. Association for Computing Machinery, New York, pp 239–249

Colon C, Hallegatte S, Rozenberg J (2021) Criticality analysis of a country's transport network via an agent-based supply chain model. Nat Sustain 4:209–215

Edwards B, McCarthy JD, Mataic DR (2018) The resource context of social movements. In: Snow DA, Soule SA, Kriesi H, McCammon HJ (eds) The Wiley Blackwell companion to social movements, 2nd edn. Wiley, Hoboken, pp 79–97

Fernandez RM, McAdam D (1988) Social networks and social movements: multiorganizational fields and recruitment to Mississippi freedom summer. Sociol Forum 3:357–382

Frank AB, Posard MN, Helmus TC, Marcinek K, Grana J, Kahn O, Zutshi R (2022) An exploratory examination of agent-based modeling for the study of social movements. RAND Corporation, Santa Monica

Geard N, Bullock S (2008) Group formation and social evolution: a computational model. Artif Life XI:197–203

Geard N, Bullock S (2010) Competition and the dynamics of group affiliation. Adv Complex Syst 13:501–517

González-Bailón S, Borge-Holthoefer J, Rivero A, Moreno Y (2011) The dynamics of protest recruitment through an online network. Sci Rep 1:1–7

Granovetter M (1973) The strength of weak ties. Am J Sociol 78:1360–1380

Hassanpour N (2010) Dynamic models of mobilization in political networks. In: Proceedings of the 2010 OpenSIUC political networks conference. Southern Illinois University Carbondale, Carbondale. http://opensiuc.lib.siu.edu/pnconfs_2010/43. Accessed 2 Aug 2024

Hassanpour N (2016) Leading from the periphery and network collective action. Cambridge University Press, New York

Hu H-H, Lin J, Cui W (2015) Cultural differences and collective action: a social network perspective. Complexity 20:68–77

Hu H-H, Wang L, Jiang L, Yang W (2019) Strong ties versus weak ties in word-of-mouth marketing. BRQ Bus Res Q 22(4):245–256

Kim J-W, Hanneman R (2011) A computational model of worker protest. J Artif Soc Soc Simul 14:1–23

Klandermans B (1986) New social movements and resource mobilization: the European and the American approach. Int J Mass Emerg Disasters 4:13–37

Knoke D (1988) Incentives in collective action organizations. Am Sociol Rev 53:311–329

Knoke D, Wisely N (1990) Social movements. In: Knoke D (ed) Political networks: the structural perspective. Cambridge University Press, New York, pp 57–84

Knoke D, Wright-Isak C (1982) Individual motives and organizational incentive systems. Res Sociol Organ 1:209–254

Larson JM, Nagler J, Ronen J, Tucker JA (2019) Social networks and protest participation: evidence from 130 million Twitter users. Am J Polit Sci 63:690–705

LeBon G (1896) The crowd. Ernest Benn, London

Marcinek K, Zutshi R, Khan O, Grana J, Posard M, Helmus T, Frank A (2022) The role of communication and network technology in the dynamics of social movements. Working paper WR-A 1646-2. RAND Corporation, Santa Monica

McAdam D (1986) Recruitment to high-risk activism: the case of freedom summer. Am J Sociol 92:64–90

McAdam D (1992) Gender as a mediator of the activist experience: the case of freedom summer. Am J Sociol 97:1211–1240

McAdam D, Paulsen R (1993) Specifying the relationship between social ties and activism. Am J Sociol 99:640–667

McCarthy JD, Zald MN (1977) Resource mobilization and social movements: a partial theory. Am J Sociol 82:1212–1241

McCarthy JD, Zald MN (2001) The enduring vitality of the resource mobilization theory of social movements. In: Turner JH (ed) Handbook of sociological theory. Kluwer Academic/Plenum Publishers, New York, pp 533–565

McCurry J, Taylor J, Roy EA, Safi M (2020) George Floyd: protests take place in cities around the world. The Guardian, 1 June

Mosler H-J, Tobias R (2001) Who participates in a collective action? A psychologically based simulation with 10,000 agents. In: 2nd workshop on agent-based simulation. SCS-Europe, Ghent, pp 77–82

Olson M (1965) The logic of collective action: public goods and the theory of groups. Harvard University Press, Cambridge, MA

Opp KD (1988) Grievances and participation in social movements. Am Sociol Rev 53:853–864

Petrov A, Akhremenko A, Zheglov S (2023) Dual identity in repressive contexts: an agent-based model of protest dynamics. Soc Sci Comput Rev 41:2249–2273

Raphiri TS, Joey J, van Vuuren J, Buitendag AAK (2023) Modeling and simulating student protests through agent-based framework. Int J Cyber Warf Terror 13:1–20

Saunders C (2022) Social networks and recruitment for environmental movements. In: Grasso M, Giugni M (eds) Routledge handbook of environmental movements. Routledge, Milton Park, pp 390–404

Snow A (2020) AP tally: arrests at widespread US protests hit 10,000. AP News, 4 June

Snow DA, Zurcher LA, Ekland-Olson S (1980) Social networks and social movements: a microstructural approach to differential recruitment. Am Sociol Rev 45:787–801

Summad E, Al-Kindi M, Al-Hinai N, Shamsuzzoha A, Piya S (2023) The application of agent-based modelling for the diffusion of innovation research: a case study. Int J Bus Innov Res 30:542–564

Taylor A (2020) Images from a worldwide protest movement. The Atlantic, 8 June

Taylor DB (2021) George Floyd protests: a timeline. New York Times, 5 November

Tindall D, Stoddart MCJ, McLevey J, Jasny L, Fisher DR, Earl J, Diani M (2021) On movements: the opportunities and challenges of studying social movement ego-networks: online and offline. In: Small ML, Perry BL, Pescosolido BA, Smith EB (eds) Personal networks: classic readings and new directions in egocentric analysis. Cambridge University Press, New York, pp 696–717

Turner RH, Killian LM (1972) Collective behavior. Prentice-Hall, Englewood Cliffs

United Nations' Post (2020) Racism is an abhorrence we must all reject. 20 June. https://www.linkedin.com/posts/united-nations_i-am-heartbroken-to-see-violence-on-the-activity-6674017020314484736-s4zV. Accessed 21 Dec 2024

Van Laer J (2017) The mobilization dropout race: interpersonal networks and motivations predicting differential recruitment in a national climate change demonstration. Mobilization 22:311–329

Walgrave S, Wouters R (2022) More than recruitment: how social ties support protest participation. Soc Probl 69:997–1024

Zald MN, McCarthy JD (2002) The resource mobilization research program: progress, challenge, and transformation. New Direct Contemp Sociol Theory:147–171

Insurrecting

Keywords Capitol Insurrection ABM · Frustration-aggression theory · Insurrection · Relative-deprivation theory

1 Introduction

On the afternoon of January 6, 2021, a large mob of insurrectionists violently attacked and breached the U.S. Capitol, intent on stopping Congress from certifying the Electoral College results of the 2020 presidential election. President Donald J. Trump, who was defeated by Joe Biden, had devoted the preceding months to spreading his big lie about a stolen election. On December 19, 2020, he tweeted, "Big protest in D.C. on January 6th. Be there, will be wild!" Thousands of his supporters came to Washington that day, many wearing body armor and carrying weapons. In a rally on the Ellipse, Trump told his followers, "If you don't fight like hell, you're not going to have a country anymore" (U.S. House of Representatives 2022: 72). He urged the crowd to walk to the Capitol and promised to meet them there:

> Now, it is up to Congress to confront this egregious assault on our democracy. And after this, we're going to walk down, and I'll be there with you, we're going to walk down, we're going to walk down. Anyone you want, but I think right here, we're going to walk down to the Capitol, and we're going to cheer on our brave senators and congressmen and women, and we're probably not going to be cheering so much for some of them. (U.S. House 2022: 73)

The original version of the chapter has been revised. A correction to this chapter can be found at https://doi.org/10.1007/978-3-031-86199-4_10

Supplementary Information The online version contains supplementary material available at https://doi.org/10.1007/978-3-031-86199-4_7.

D. Knoke, *Network Collective Action*, Lecture Notes in Social Networks, https://doi.org/10.1007/978-3-031-86199-4_7

Trump's Secret Service detail refused to drive him to the Capitol, instead taking him back to the White House where he watched the attack on television. He ignored pleas from his family, members of his administration, and Republican elected officials to order the mob to go home.

Thousands of people streamed down Pennsylvania Avenue, arriving at 1:10 p.m. outside the Capitol building where they proceeded to beak barricades, scale walls, smash windows and doors, assault Capitol police officers, trash offices, and steal property, while chanting "Hang Mike Pence" because the vice president had refused to reject the certified electors. More than 140 police officers were injured, four died of natural causes during the attack, and four died later by suicide. "A woman who attempted to forcibly enter the Chamber of the House of Representatives through a broken window while the House was in session was shot and killed by police guarding the chamber" (U.S. House 2022: 76). More than 2000 people gained access to the interior of the Capitol. Secret Service professionals rushed the vice president, his family, and other legislators to secure locations. Trump finally released a video at 4:17 p.m. telling his supporters, "We have to have peace. So go home. We love you. You're very special" (U.S. House 2022: 580). The D.C. National Guard was deployed to clear and secure the building. Pence gaveled the Senate back into session, as did Speaker Nancy Pelosi in the House. At 3:24 a.m. on January 7, a majority of Congress voted to confirm Biden's electoral college victory, although 127 Republican legislators backed objections to certifying Arizona's electoral outcome and 145 Republicans supported objections to certifying Pennsylvania's results (Zhou 2021).

Many Capitol attackers were unconnected to extremist groups, but the most violent attackers belonged to organized militant networks, such as the Proud Boys, Oath Keepers, and Three Percenters. Their participation in the attack was carefully planned and coordinated through network relations:

> Militant networks at the Capitol were characterized by hierarchical organization and chains of command. Leaders of established domestic violent extremist groups issued orders or directives to members of their groups, encouraging them to travel to Washington in advance of the siege. Individual group members answered the call, contacting one another to coordinate logistics, methods, and plans of action in the weeks before January 6th. Unlike individuals in the other categories, not only did these militant networks plan to attend protests on the 6th, but they are also alleged to have planned in advance to breach the Capitol and, in many cases, conduct violence inside the walls of the building. (Vidino et al. 2021: 17)

Data assembled from various sources by the University of Maryland's National Consortium for the Study of Terrorism and Responses to Terrorism (START), showed about one-third of the initial 800 individuals charged with crimes had preexisting ties to extremist groups. A network graph revealed that Proud Boys leader Enrique Tarrio and Oath Keepers leader Stewart Rhodes "connect almost three-quarters of the known extremist organizations that were present on Jan. 6" (Rogers et al. 2022). In the 3 years following the attack, Department of Justice prosecutors charged more than 1230 participants with federal crimes "ranging from misdemeanors like trespassing to felonies like assaulting police officers and seditious conspiracy" (Richer and Kunzelman 2024). Approximately 730 people pleaded guilty to charges and 170 were convicted at a trial decided by a judge or jury. Tarrio and Rhodes, along with a half-dozen others, were tried for seditious conspiracy. Both were convicted and sentenced to 22 and 18 years' incarceration, respectively.

The Capitol attack is often described as an insurrection (e.g., Hinsz and Jackson 2022; Holt 2022). The 18 U.S. Code 2382 Rebellion or Insurrection states, "Whoever incites, sets on foot, assists, or engages in any rebellion or insurrection against the authority of the United States or the laws thereof, or gives aid or comfort thereto shall be fined under this title or imprisoned not more than ten years, or both; and shall be incapable of holding any office under the United States." A week after the attack, the US House impeached Trump for a second time, charging him with "incitement of insurrection," because he "willfully made statements that encouraged – and foreseeably resulted in – imminent lawless action at the Capitol." At the Senate trial the ex-President was acquitted when the tally to convict came 10 votes short of the two-thirds majority required by the Constitution.

Insurrections are a type of collective action that fall on an institutional spectrum between the unorganized riots in chapter "Rioting" and the insurgents and guerrillas fighting the police and armed forces of a government in chapter "Insurging". The next section discusses theories and research on political violence, including applications to the Capitol attack. The following section examines some agent-based models of insurrectionary attacks on government buildings, forts, camps, and similar facilities. The next section presents an ABM simulating the January 6 Capitol insurrection. The chapter concludes with some suggestions for future directions of research and theory development.

2 Theories and Research on Political Violence

This section discusses two closely related theories about the micro-level mechanisms that explain why people engage in aggressive and violent acts: frustration-aggression (F-A) and relative deprivation (RD). It discusses how these theories explain participation in political violence, defined as "all collective attacks within a political community against the political regime, its actors – including competing political groups as well as incumbents – or its policies. The concept represents a set of events, a common property of which is the actual or threatened use of violence" (Gurr 1970: 3). It examines a few applications of those two theories to recent instances of political violence, including the Capitol insurrection.

A durable explanation of political violence is the frustration-aggression hypothesis. In its original formulation by Yale University psychologists, "the occurrence of aggressive behavior always presupposes the existence of frustration and, contrariwise, that the existence of frustration always leads to some form of aggression" (Dollard et al. 1939: 1). Aggressive responses may range from threats and verbal abuse to physical assault and even homicide. The target of aggression could be the source of the frustration, or displaced onto a scapegoat. For example, a worker denied a promotion would not curse her manager for fear of being fired, but instead go home and rant to her spouse and children. Frustration is "an interference with the occurrence of an instigated goal-response" (Dollard et al. 1939: 7). A frustrating event or life situation blocks a person from obtaining an important goal, which then inevitably results some type of aggressive behavior. The authors subsequently modified their hypothesis to remove the certainty of aggression: "frustration produces instigation to aggression but this is not the only type of instigation that it may produce" (Miller et al. 1941: 339). Leonard Berkowitz (1989: 71) offered a major modification of F-A theory: "Frustrations are aversive events and generate aggressive inclinations only to the extent that they produce negative affect." Frustrations are only one of many potential sources of negative affect, thus making frustration a sufficient rather than a necessary cause of aggression. Decades of F-A research on individuals and small groups developed more nuanced understanding of factors that mediate, moderate, and condition the link connecting frustration to aggression. For example, frustration must represent a loss of significance and "the primordial impulse to reassert one's significance by proving one's power and dominance must be more salient and accessible to the individual than alternative nonaggressive means to significance, channeled via values other than power and dominance" (Kruglanski et al. 2023: 462). A review of research by Johannes Breuer and Malte Elson (2017: 9) concluded that F-A theory "continues to be a valuable asset in the work of psychologists and other social scientists interested in the study of human aggression."

In *The American Soldier*, Samuel Stouffer et al. (1949) proposed relative deprivation theory to explain such apparent paradoxes as African American soldiers in Southern camps being more satisfied than those in Northern camps. RD can be defined "as a judgment that one or one's ingroup is disadvantaged compared to a relevant referent and that this judgment invokes feelings of anger, resentment, and entitlement" (Smith and Pettigrew 2015: 2). The concept refers to individuals and their referents, which may be individuals or groups.

> Individuals undergoing RD experience in turn three psychological processes: (1) they first make cognitive comparisons, (2) then cognitive appraisals that they or their ingroup are disadvantaged, and finally (3) that these disadvantages are seen as unfair and arouse angry resentment. (Pettigrew 2015: 12)

The basis of comparison may be various desired social resources, such as money, justice, civil rights, status, and honor. People must perceive the disadvantage as unfair, undeserved, or discriminatory. Researchers demonstrated that RD affects a wide variety of beliefs, emotions, and behaviors, such as despair, resentment, prejudice, scapegoating, anger, unhappiness, depression, ill health, or criminal activity. RD may dispose people to embrace political ideologies that explain the causes of the deprivation and endeavor to reduce or eliminate those causes. Group RD is a motivator of liberal collective action by deprived people to change their group's disadvantages (Osborne and Sibley 2015; Lilly et al. 2022). In contrast, people who believe inequality is acceptable are prone to embrace conservative ideologies that oppose collective actions aimed at changing the status quo.

Several analysts found evidence consistent with F-A or RD hypotheses. Using surveys of United Kingdom voters in the 2017 General Election and 2019 European parliamentary elections, Sarah Harrison (2020) found that the political and institutional dimensions of frustration combined to lead citizens to consider taking part in violent demonstrations or even joining a revolution. "When citizens' frustration is based on the desire for both functioning institutions and decent politicians to be failed by major perceived delivery gaps is when they consider all other options but aggression to potentially having been exhausted" (p. 25). Following three of the deadliest U.S. mass shootings, researchers conducted four studies on 2442 gun owners to test hypotheses about "whether mass shooting salience interacted with thwarted goals to predict justification to shoot suspected criminals, as well as ideas about armed vigilantism and perceptions that guns are means of empowerment" (Leander et al. 2019: 704). Thwarted goals (i.e., frustration) were experimentally induced in one study by failure on an achievement task and in three studies by perceptions of disempowerment in society. The results revealed a positive relation between frustration and support for gun violence.

Analyzing two samples of Muslim youths living in Denmark, Milan Obaidi et al. (2018) found that perceived group deprivation, but not individual deprivation, predicted greater intentions to commit acts of violence in Europe. Western-born Muslims without personal experience of adversity exhibited "victimization-by-proxy," showing highly similar relative deprivation to those foreign-born youths with adverse experiences of war and occupation in the Middle East. "This finding is crucial as it highlights the importance of understanding violent extremist phenomena from a transnational perspective, rather than viewing incidents as single, isolated events" (Kunst and Obaidi 2020: 58). Glenn Starks (2022) investigated the impact of relative deprivation on antithetical movements opposed to the Black Lives Matter movement against racism. White supremacists seeking to restore their political dominance perceive structural disadvantages and loss of power as unfair, arousing angry resentment. "This angry resentment is then turned into action through the use of symbolic racism, where antithetical movements use symbols of the confederacy, racial epithets, and distorted patriotism as part of their rallying cries" (p. 353). In an online survey of 455 Americans about the Capitol attack, Tanner Butler (2022: 8) found an extremely strong connection "between respondents' support for the riot and their belief that the rioters were within their right to rebel." He speculated the respondents who justified the insurrection were implicitly agreeing with English Enlightenment philosopher John Locke's theory that the people have a natural right to overthrow a tyrannical government:

> While respondents may not have strictly adhered to what Locke meant, the basic ideas are still there. That is what is important for this relationship. It is not just anger and relative deprivation surrounding the 2020 Presidential election causing support. Since respondents say that protestors are reaching the point that natural rights are under threat. So, we can see there is something deeper going on. The reason why respondents view the results as affecting their natural rights is unknown and needs further study. Without this, we can tell the underlying cause of the relative deprivation is salient to respondents; since they indicated it was bad enough that the protestors were within their right to overthrow the government. (Butler 2022: 11)

The micro-mechanisms of political violence inform the construction of agent-based models discussed in the following sections.

3 Political Violence ABMs

Numerous researchers have developed agent-based models of contentious politics, social conflict, and political violence (e.g., Lemos 2018; Deutschmann et al. 2020; Dacrema and Benati 2020; Nurdiansyah et al. 2023; Macmillan-Scott et al. 2024).

Many ABMs investigated multiple encounters between militants and governmental agents in different locations over time, the topic of chapter "Insurging". Some recent models focused a single event occurring at one location where protestors clash with police, the topic of this chapter. Seyed Hozhabrossadati (2022) focused on micro-level modeling of protestor movements toward a facility guarded by police who have authority to arrest people entering an "alert area." Relative deprivation was operationalized as the difference between each protestor's wage and the most highly paid agent in a circular neighborhood. He concluded that "rioters are willing to reach valuable sites while escaping from colliding obstacles and being arrested by policemen. ... Results also show that legitimacy can play an important role in crowd control; the higher the government legitimacy, the lower the grievance" (Hozhabrossadati 2022: 68)

Tshepo Raphiri et al. (2023) constructed an ABM of student protests at South African universities. Their main focus was "to evaluate how the degree of inequality level, number of activists, activists' influential size, number of friendship ties, suspend delay and sympathy affect the dynamics of student protests" (p. 12). Suspend delay reduces the risk of activist students being suspended by law enforcement officers. Based on experiments with varied parameters settings, the authors concluded that "when these independent variables have increased in various scenarios studied, both the volume of outbursts and strength of protests increases" (p. 17). Based on 20 ethnographic qualitative interviews in Oaxaca, Mexico, Travis Holmes (2023) constructed an ABM of the 2006 teachers union occupation of the downtown public square and their violent removal by police, resulting in many arrests and more than 100 people hospitalized with injuries. The outcomes of seven scenarios with varied parameter values, he analyzed the impact of police and information on protest emergence, sustainability, and termination. For example, "police presence, in absolute terms (i.e., the literal number of police in the vicinity of the protests), does make a difference when it comes to the average protest level" (p. 120). However, "there is no clear answer as to precisely how many police, and under what conditions police presence can lead to termination as witnessed in Oaxaca in 2006" (p. 121).

In an ABM that very closely resembles the Capitol insurrection, Carlos Lemos et al. (2014, 2015) modelled violent confrontations during a single protest event taking place at an "attraction point" such as a government facility. The model created by Lemos et al. (2014) has three primary types of agents—protestors, cops, and media—with three subtypes of protestors have varying propensities to turn violent—passer-by, hanger-on, and hardcore. The cops try to defend the attraction point against violent protestors, keep close to one another, and if they have enough backup, to arrest violent protestors. Protestors gravitate toward media agents who "try to locate and record ('take pictures' of) violent events" (p. 41). The ABM program involves protestors scanning, planning, and acting—transitioning among quiet, active, and violent states—according to a variant of Joshua Epstein's threshold rule (see chapter "Insurging"). Transitions occur when the level of grievance is greater than the perception of net risk. The researchers ran simulations with varying parameters on a scenario where protestors try to reach the entrance to a government building, such as the Parliament in Lisbon, Portugal, and are opposed by a police force. The results "described several emergent crowd patterns in real protests, such as clustering of violent and active protestors and formation of a confrontation line moving back and forth with localized fights. Violent behavior was restricted to the initially more aggressive protestors and did not propagate to the bulk of the crowd" (Lemos et al. 2014:136).

4 Capitol Insurrection ABM

"Capitol Insurrection" is a NetLogo ABM that captures some key features of the January 6, 2021, assault on the US Capitol building by a mob of protestors trying to prevent Congress from certifying the results of the 2020 Presidential election. The top of the world is a schematic of the Capitol building with the Senate on the left and House on the right. Two imaginary interior rooms are a brig for arrested insurrectionists and an infirmary for injured police officers. Two entrances to the building, a door on the left and a window on the right, are initially closed. At bottom of the world is a plaza where bystanders and insurrectionists initially assemble. The names of the five types of agents, their colors, and behaviors are:

(1) **Bystanders** (orange) are nonviolent protestors who randomly mill around the plaza without intent to enter the Capitol. When a bystander encounters two insurrectionists and its threshold for recruitment is exceeded, the bystander becomes a recruit able to enter the building.

(2) **Recruits** (green) are bystanders recruited by militants. They change color from orange to green and enter the building. They do not try to enter the Senate or House but stay in the atrium between those chambers.

(3) **Proud Boys** (red) are one of the two insurrectionist organizations. They seek to enter the Capitol and occupy the Senate. Entering the building requires that two Proud Boys smash the door.

(4) **Oath Keepers** (red) are the second militant group. They seek to enter the Capitol and occupy the House. Entering the building requires that two Oath Keepers break the window.

(5) **Capitol Police** (blue) are located inside the Capitol building where they attempt to arrest Proud Boys and Oath Keepers who enter the building. If they succeed in arresting a militant, they're sent to the brig. In turn, the militants violently attack police who, when injured, are taken to the infirmary.

Sliders on the interface allow the researcher to choose the numbers of agents and the probabilities for bystander recruitment, arrests of militants, and injuries to police. The simulation ends when all insurrectionists either enter one of the chambers or are sent to the brig. A plot tracks over-time changes in the percentage of agents in the chambers, brig, and infirmary while monitors display the tallies at the end of the simulation.

Two screen captures demonstrate simulation outcomes with differing numbers of insurrectionists and police while keeping the probabilities of recruitment, injury, and arrest at 0.50 in both experiments. In Fig. 1, the 90 insurrectionists outnumber

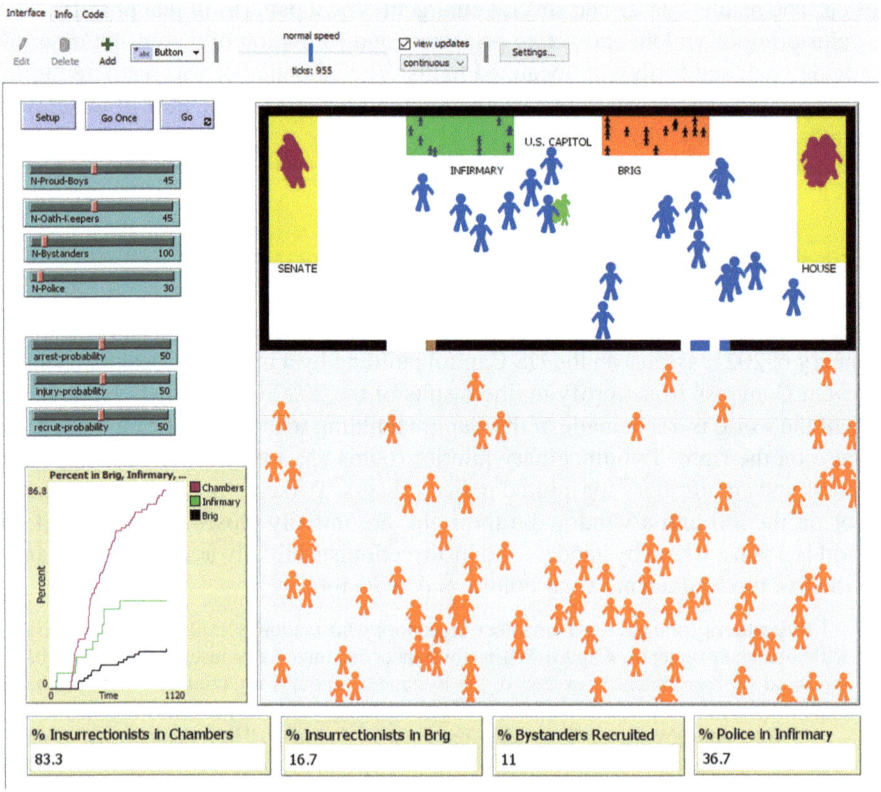

Fig. 1 Capitol Insurrection ABM with insurrectionists outnumbering police

30 police. As a result, 83.3% of insurrectionists succeed in occupying the two chambers, while 36.7% of the police are injured. Figure 2 reverses the sizes of the opposing forces: 150 police outnumber 30 insurrectionists. Consequently, 83.3% of the insurrectionists are arrested and only 7.3% of the police are injured. Those two outcomes were cherry picked to exaggerate the contrasting outcomes of imbalanced opposing forces. Table 1 shows the results after running each experiment 200 times. When the police are outnumbered, the mean percentage of insurrections occupying the two chambers is 82.2% and 18.2% of police are injured. When the police outnumber the insurrectionists, the mean percentage of insurrections arrested is 74.0% and 7.4% of police are injured. The clear result is than when police are overwhelming the dominant force, they prevent most insurrectionists from breaching the Capitol chambers while suffering fewer injuries.

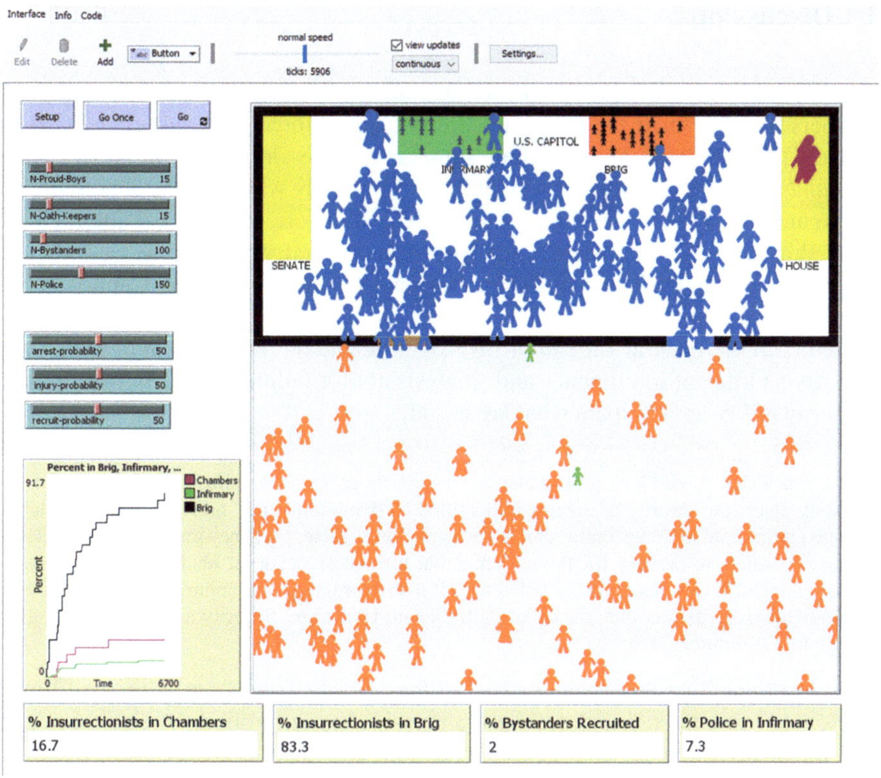

Fig. 2 Capitol Insurrection ABM with police outnumbering insurrectionists

Table 1 Two experiments with different ratios of insurrectionists to police

	Model 7.1	Model 7.2
Number of insurrectionists	90	30
Number of police	30	150
Mean % insurrectionists in chambers	82.2%	26.0%
Mean % insurrectionists arrested	17.8%	74.0%
Mean % police injured	18.2%	7.4%
Mean ticks	2396	5466
N of experimental runs	200	200

5 Discussion

The Capitol insurrectionists succeeded in entering the building because their sheer numbers overwhelmed the law enforcement counterforce. The subtitle of Capitol Police Chief Steven Sund's memoir summarized the situation: *Under Siege and Outnumbered 58 to 1 on January 6* (2023). The tragedy was a failure of both imagination and intelligence. The U.S. House (2022: 6) asserted that "the Capitol Police leadership did not anticipate the scale of violence that would ensue ... Although Chief Steven Sund raised the idea of National Guard support, the Capitol Police Board did not request Guard assistance prior to January 6th." A Senate investigation faulted both the Federal Bureau of Investigation and the Department of Homeland Security's Office of Intelligence and Analysis not for failures to obtain intelligence, but for inability to anticipate what lay ahead:

> [They] obtained multiple tips from numerous sources in the days and weeks leading up to the attack that should have raised alarms. Rather, those agencies failed to fully and accurately assess the severity of the threat identified by that intelligence, and formally disseminate guidance to their law enforcement partners with sufficient urgency and alarm to enable those partners to prepare for the violence that ultimately occurred on January 6th. At a fundamental level, the agencies failed to fulfill their mission and connect the public and nonpublic information they received. (U.S. Senate Homeland Security and Governmental Affairs Committee 2023: 3)

If the National Guard had been called out long before Trump's rally on the Ellipse, they could have erected and defended a massive barrier around the Capitol to prevent the insurrectionists from getting anywhere near their target.

The Capitol insurrection was an ineffective one-day revolt against the national government in support of a soon-to-be ex-president's big lie about a stolen election. In the present century, several dozen more sustained insurrections may be concurrently under way around the world (Barria-Asenjo et al. 2022). A precise number is indeterminate due to the wide range of labels and definitions applied used by scholars and policy makers. When do protests swell into uprisings, insurrections balloon into insurgencies, or rebellions leap into revolutions? Such ambiguities should not hinder the development of more useful models to illuminate the origins of violent civil conflicts, their convoluted dynamics, and their consequential outcomes.

References

Barria-Asenjo NA, Žižek S, Scholten H, Pavón-Cuellar D, Salas G, Cabeza OA, Arohuanca JWH, Aguilar SJ, Alcalá. (2022) Returning to the past to rethink socio-political antagonisms: mapping today's situation in regards to popular insurrections. CLCWeb Comp Lit and Cult 24:1–15

Berkowitz L (1989) Frustration-aggression hypothesis: examination and reformulation. Psychol Bull 106:59–73

Breuer J, Elson M (2017) Frustration-aggression theory. In: Sturmey P (ed) The Wiley handbook of violence and aggression. Wiley Blackwell, Chichester, pp 1–12

Butler T (2022) The right to rebel and the insurrection at the Capitol: what causes support for the events of January 6th. Undergrad Res J 26(Article 2):1–16
Dacrema E, Benati S (2020) The mechanics of contentious politics: an agent-based modeling approach. J Math Sociol 44:163–198
Deutschmann E, Lorenz J, Nardin LG, Natalini D, Wilhelm AFX (eds) (2020) Computational conflict research. Springer Nature, Berlin
Dollard J, Miller NE, Doob LW, Mowrer OH, Sears RR (1939) Frustration and aggression. Yale University Press, New Haven
Gurr TR (1970) Why men rebel. Princeton University Press, Princeton
Harrison S (2020) Democratic frustration: concept, dimensions and behavioural consequences. Societies 10:19–30
Hinsz VB, Jackson JW (2022) The relevance of group dynamics for understanding the U.S. Capitol insurrection. Group Dyn Theory Res Pract 26:169–177
Holmes T (2023) Complex dynamics of contention: towards a generative model of social dissent. Doctoral Dissertation, Department of Political Science and Geography. Old Dominion University, Norfolk
Holt J (2022) After the insurrection. Atlantic Council, Washington, DC. www.atlanticcouncil.org/wp-content/uploads/2022/01/After-the-Insurrection.pdf. Accessed 12 June 2024
Hozhabrossadati SM (2022) Simulating crowd behavior using artificial potential fields: an agent-based simulation approach. J Syst Think Pract 1:49–71
Kruglanski AW, Ellenberg M, Szumowska E, Molinario E, Anne Speckhard N, Leander P, Pierro A, Di Cicco G, Bushman BJ (2023) Frustration-aggression hypothesis reconsidered: the role of significance quest. Aggress Behav 49:445–468
Kunst JR, Obaidi M (2020) Understanding violent extremism in the 21st century: the (re) emerging role of relative deprivation. Curr Opin Psychol 35:55–59
Leander NP, Stroebe W, Kreienkamp J, Agostini M, Gordijn E, Kruglanski AW (2019) Mass shootings and the salience of guns as means of compensation for thwarted goals. J Pers Soc Psychol 116:704–723
Lemos CM (2018) On agent-based modeling of social conflict: from mechanisms to complex behavior. Springer Nature, Cham
Lemos C, Coelho H, Lopes RJ (2014) Agent-based modeling of protests and violent confrontation: a micro-situational, multi-player, contextual rule-based approach. In: Proceedings of the 5th World congress on social simulation Sao Paulo, Brazil, 4–5 November, pp 136–160
Lemos C, Lopes RJ, Coelho H (2015) Quantitative measures of crowd patterns in agent-based models of street protests. In: Proceedings of 2015 IEEE World Conference on Complex Systems, WCCS 2015. IEEE, Marrakesh, pp 1–6
Lilly KJ, Sibley CG, Osborne D (2022) Different domains of area-level deprivation predict individual differences in system justification and collective action support. N Z J Psychol 51
Macmillan-Scott O, Ünver A, Musolesi M (2024) Game-theoretic agent-based modelling of micro-level conflict: evidence from the ISIS-Kurdish war. PLoS One 19(6):e0297483
Miller NE, Sears RR, Mowrer OH, Doob LW, Dollard J (1941) The frustration aggression hypothesis. Psychol Rev 48:337–342
Nurdiansyah H, Almubaroq HZ, Risdhianto A, Mualim M (2023) Evaluation of the spread of radicalism, extremism, and terrorism in Indonesia's defense using agent-based simulations. Int J Humanit Stud 6:228–239
Obaidi M, Bergh R, Sidanius J, Thomsen L (2018) The mistreatment of my people: victimization by proxy and behavioral intentions to commit violence among Muslims in Denmark. Polit Psychol 39:577–593
Osborne D, Sibley CG (2015) Opposing paths to ideology: group-based relative deprivation predicts conservatism through warmth toward ingroup and outgroup members. Soc Justice Res 28:27–51
Pettigrew TF (2015) Samuel Stouffer and relative deprivation. Soc Psychol Q 78:7–24

Raphiri TS, Joey J, van Vuuren J, Buitendag AAK (2023) Modeling and simulating student protests through agent-based framework. Int J Cyber Warf Terror 13:1–20

Richer AD, Kunzelman M (2024) Hundreds of convictions, but a major mystery is still unsolved 3 years after the Jan. 6 Capitol riot. Associated Press, 5 January. https://apnews.com/article/capitol-riot-jan-6-criminal-cases-anniversary. Accessed 23 Dec 2024

Rogers K, Fuong H, Mejia E, Sweedler M, Matlin C (2022) Jan. 6's tangled web of extremism. FiveThirtyEight.com, 11 July. https://fivethirtyeight.com/features/jan-6s-tangled-web-of-extremism. Accessed 12 June 2024

Smith HJ, Pettigrew TF (2015) Advances in relative deprivation theory and research. Soc Justice Res 28:1–6

Starks GL (2022) Explaining antithetical movements to the Black Lives Matter movement based on relative deprivation theory. J Black Stud 53:346–365

Stouffer SA, Suchman EA, DeVinney LC, Starr SA, Williams RM (1949) The American soldier: adjustment to army life, vol Volume 1. Princeton University Press, Princeton

Sund SA (2023) Courage under fire: under siege and outnumbered 58 to 1 on January 6. Blackstone Publishing, Ashland

U.S. House of Representatives (2022) Final report select committee to investigate the January 6th attack on the United States Capitol. December 22, 2022. 117th Congress Second Session House Report 117-663. Accessed 12 June 2024

U.S. Senate Homeland Security and Governmental Affairs Committee (2023) Planned in plain sight: a review of the intelligence failures in advance of January 6th, 2021. Accessed 8 Aug 2024

Vidino L, Hughes S, Meleagrou-Hitchens A, Margolin D, Clifford B, Lewis J, Mines A, Ingram H (2021) 'This is our house!' A preliminary assessment of the Capitol Hill siege participants. Program on Extremism at George Washington University, Washington. https://extremism.gwu.edu/sites/g/files/zaxdzs5746/files/This-Is-Our-House.pdf. Accessed 12 June 2024

Zhou L (2021) 147 Republican lawmakers still objected to the election results after the Capitol attack. Vox. www.vox.com/2021/1/6/22218058/republicans-objections-election-results. Accessed 12 June 2024

Insurging

Yicheng Shen

Keywords Counterinsurgency · Guerilla warfare · Insurgency · Insurgents and Soldiers Fighting ABM · Revolution

1 Introduction

In the 1950s, insurgents fought to overthrow the corrupt dictator of Cuba, Fulgencio Batista (see e.g., Chomsky 2015; Pérez-Stable 2011). On July 26, 1953, a failed attack on the Moncada Barracks at Santiago de Cuba resulted in Fidel Castro sentenced to a 15-year prison term and his younger brother Raúl to a 13-year term. Believing the Castro brothers were a spent force, Batista pardoned them in May 1955. Under a government crackdown on dissidents, they fled within weeks to Mexico where they met Ernesto "Che" Guevara, an Argentine physician and leading theorist of guerrilla warfare. The insurgency was renamed MR-26-7 (*Movimiento Revolucionario 26 de Julio*) after the date of the failed barracks assault.

In late November 1956, Fidel and 81 rebels set sail from Mexico on the decrepit yacht *Granma*. Delayed by bad weather and a leaky vessel, they failed to reach Cuba in time to coordinate with urban uprisings by MR-26-7 forces. On December 2, the *Granma* crashed into a mangrove swamp at Playa Los Cayuelos in southeastern Cuba, where Batista's soldiers, helicopters, and airplanes harassed the invaders as they struggled to move inland. Only 21 insurgents escaped death or capture, including Fidel, Raúl, Che, and Camilo Cienfuegos. Reaching the safety of the Sierra Maestra mountain

The original version of the chapter has been revised. A correction to this chapter can be found at https://doi.org/10.1007/978-3-031-86199-4_10

Supplementary Information The online version contains supplementary material available at https://doi.org/10.1007/978-3-031-86199-4_8.

D. Knoke, *Network Collective Action*, Lecture Notes in Social Networks, https://doi.org/10.1007/978-3-031-86199-4_8

jungles along the coast of Oriente Province, the survivors set about expanding the insurgency. They used classic guerrilla warfare hit-and-run tactics, raiding small army posts for weapons. Recruiting students and workers from urban areas as well as local peasants, MR-26-7 eventually grew to 200 fighters by 1957. Fidel divided the forces into three columns, assigning the second to his brother and the third to Guevara. Increased attacks forced Batista's army out of the Sierra Maestra mountains. By 1958, the insurgents were operating schools, a hospital, printing press, and land-mine factory.

In June 1958, Batista launched a brutal counterinsurgency with aerial bombardment of the jungle while 10,000 soldiers surrounded the Sierra Maestra and penetrated insurgent strongholds. Their unprovoked violence against rural villages radicalized a terrorized peasantry and drove many to join the insurgency. Now commanding 300 fighters, the rebels deployed land mines and ambushes to inflict heavy casualties on the government forces. By summer, MR-26-7 again forced the army out of the Sierra Maestra and encircled Santiago and Santa Clara. Batista fled to the Dominican Republic on December 31, 1958. The insurgency came to a successful conclusion on January 2, 1959, as Guevara and Cienfuegos marched their columns into the Cuban capital Havana and Fidel accepted the surrender of the Moncada Barracks in Santiago de Cuba.

This chapter discusses theories of insurgency and counterinsurgency, reviews agent-based models of combat between insurgent and government forces, and presents and ABM of insurgents fight soldiers to seize control of jungle terrain. It concludes with suggestions for extensions of the basic model and speculations about the impact of military technology on the future shape of insurgencies.

2 Insurgency and Counterinsurgency

Insurgency is sometimes treated as synonymous with guerrilla warfare, while other analysts see the terms as conceptually distinct (e.g., Boot 2013; Hobson and Moghadam 2024). To avoid this fruitless debate, the preferred term in this chapter is insurgency, used interchangeably with guerrilla warfare. The US armed forces doctrine on counterinsurgency defines insurgency as:

> … the organized use of subversion and violence to seize, nullify, or challenge political control of a region. An insurgency is a form of intrastate conflict, and counterinsurgency (COIN) is used to counter it. The term insurgency can also refer to the group itself. Insurgents can combine the use of terrorism; subversion; sabotage; other political, economic, and psychological activities; and armed conflict to achieve its aims. It is an organization political-military struggle by a predominantly indigenous group or movement designed to weaken, subvert, or displace the control of an established government for a particular region. (U.S. Joint Chiefs of Staff 2018: I-1)

A major distinction between insurgent and terrorist organizations, such as Al-Qaida and the Irish Republican Army, is that the latter seldom have sufficient military forces to seize and hold a territory and assert sovereignty over its population (Ünal 2016; Hobson and Moghadam 2024). In contrast, insurgents try to mobilize sufficient numbers of fighters to engage in force-on-force attacks against defended government targets, with the ultimate goal of toppling and replacing the regime.

Two of the foremost communist theorists of insurgency were also prominent practitioners—Mao Zedong and Che Guevara. Each wrote a how-to pamphlet with similar titles, *On Guerilla Warfare* (Mao 1937) and *La guerra de guerrillas* (Guevara 1960). Mao helped to create the Chinese Red Army and led it on the 1934–1935 Long March to avoid encirclement by the Kuomintang National Army. During the Second World War, the Red Army put up stiff resistance to the Japanese invaders, then expelled the Kuomintang out of mainland China onto the island of Taiwan in 1949. Mao summarized military tactics in an aphorism: "The enemy advances, we retreat; the enemy camps, we harass; the enemy tires, we attack; the enemy retreats, we pursue" (Mao 1930). He acknowledged that, although military strategies and tactics are crucial, political skills are indispensable for a successful insurgency. Gaining the active support of the population holds the key to victory. In another Mao aphorism, "The guerrilla must move amongst the people as a fish swims in the sea." To achieve this objective, civilians must be treated with respect. "Mao gives three rules and eight remarks to guide guerrilla forces. Some are practical, for instance, 'do not steal from the people,' 'replace the door when you leave the house,' and 'return what you borrow'" (Fonay 2013: 3). As a Marxist theorist, he emphasized that raising class consciousness is necessary for successful outcomes. He emphasized the integration of a hierarchical military command structure with a political hierarchy. "This military-political structure mobilizes and indoctrinates the local populace in the goals of the revolution" (Knoke 2013: 2).

Drawing from his experiences in the Cuban Revolution, Guevara's "foco" theory (*foquismo*) asserted that military action alone suffices to spark class consciousness. "It is not necessary to wait until all conditions for making revolution exist; the insurrection can create them" (Guevara 1960: 1). No need to work patiently to create a class-conscious urban proletariat when immediate action can be launched by an elite guerrilla force.

> Rather than waiting for a working class to emerge and gain revolutionary ideals, rural peasants can be elicited by the guerrilla foco for support, first by hiding and informing the fighters, then joining in the foco in militant action. Peasant support can be further secured, according to Guevara, with promises of agrarian reform, which should form the staple political discourse of the guerrilla foco. (Johnson 2006: 27)

Guerrilla leaders operating in rural areas of Latin America "would advance as the nucleus of revolutionary resistance against exploitation and repression, and eventually would become the vanguard of fundamental economic and social transformation in a later phase" (Kruijt 2023: 1529). In contrast to Mao's successful campaigns against the Japanese and Kuomintang Nationalists, Guevara's record was mixed. After helping to consolidate the Cuban Revolution, he embarked on three failed African and Latin American insurgencies, ending in 1967 in Bolivia with his capture and summary execution by counterinsurgency forces advised by the US Central Intelligence Agency. Due to different sociopolitical conditions in Bolivia compared to Cuba, the guerrilla foco could not spark a revolution (Childs 1995; Johnson 2006; Kruijt 2023). The durability of China's communist revolution and the repeated failures of foco insurgencies in Latin American and Caribbean nations suggests that Mao developed the more useful theory of insurgency.

Maoist-style insurgencies begat counterinsurgency strategies that "aimed to decapitate an insurgency's politico-military leaders and to dry up the sea of the people (a.k.a., winning-hearts-and-minds)" (Knoke 2013: 2). Although the British achieved notable success during the Malayan Emergency in the 1950s, US counterinsurgency efforts in Vietnam in the 1950s and 1960s failed to stem the eventual victory of the Viet Cong and North Vietnamese army. During the Iraq War (2003–2011) and the War in Afghanistan (2001–2021), the US and Coalition forces in alliance with the national governments faced prolonged insurgencies aimed at expelling the occupiers and ousting the collaborative regimes. By late 2004, a group of soldiers, academics, and policy makers began crafting new counterinsurgency strategies and tactics, culled from lessons learned from some historical COIN successes and failures, including Malaya and Vietnam. Army General David Petraeus and Marine General James Amos co-authored the *U.S. Army/Marine Counterinsurgency Field Manual 3-24* (2006). They endorsed the hearts-and-minds approach, bluntly declaring in their Foreword:

> Soldiers and Marines are expected to be nation builders as well as warriors. They must be prepared to help reestablish institutions and local security forces and assist in rebuilding infrastructure and basic services. They must be able to facilitate establishing local governance and the rule of law.

Near the end of the manual, they characterized successful counterinsurgency as "armed social work" (p. 299). Several of their doctrine's core principles strikingly mirrored Mao's requirements for successful insurgencies.

- Protect the population by understanding its culture, ideology, religion
- Learn how the populace interacts with insurgents, nongovernmental organizations, governments
- Use public diplomacy to render insurgents ineffective and noninfluential
- Political, social, economic programs are often more valuable than conventional military ops in defeating insurgencies
- Apply social network analysis to gather and interpret intelligence on insurgents

By living among the local residents instead of remote fortified camps, COIN units could use social networking to gain trust. Those connections would give soldiers access to timely intelligence on the insurgents' whereabouts, arms cache locations, and planned attacks. As the commander of Multinational Force Iraq during the 2007 troop surge, Petraeus credited COIN with turning around the Iraq War, leading to the eventual withdrawal of all occupiers.

Afghanistan was a different story. General Stanley McChrystal applied COIN doctrine, especially social networking: "In bitter, bloody fights in both Afghanistan and Iraq, it became clear to me and many others that to defeat a networked enemy we had to become a network ourselves" (McChrystal 2011: 2). In 2009, McChrystal asked for 40,000 more troops, but President Barack Obama sent only 30,000 and set 18 months as the deadline for their withdrawal. "Under those resource constraints, McChrystal's effort to implement COIN ultimately failed. Some U.S. field commanders simply ignored orders to protect civilians, giving top priority instead to conventional search-and-destroy operations against the Taliban" (Knoke 2013: 6). For decades to come, scholars and policymakers will debate the reasons for COIN failure and the Taliban's eventual reconquest of Afghanistan.

3 ABMs of Insurgencies & Counterinsurgencies

This section reviews three agent-based models depicting conflicts between opposing forces. A primary objective is to explain the conditions under which one side or the other is ultimately victorious.

Civil Violence Model Joshua Epstein proposed a simple agent-based model of spontaneous civil violence by an oppressed population against a government. The first of two variants has only two types of agents. "In the first a central authority seeks to suppress decentralized rebellion. In the second a central authority seeks to suppress communal violence between two warring ethnic groups" (Epstein 2002: 7243). In the primary version, agents are members of the general population that may or may not be actively rebellious. Cops seek out and arrest actively rebellious agents. An agent's grievance is the product of perceived hardship (physical or economic privation) and perceived illegitimacy of the regime. Risk aversion is the estimated likelihood of arrest prior to joining a rebellion. This estimate increases with the ratio of cops to actively rebellious agents within a prospective rebel's neighborhood. In effect, an agent asks, "How likely am I to be arrested if I go active?" (p. 7243). A cop's behavior is determined by the rule "Inspect all sites in the neighborhood and arrest a random active agent. Arrested rebels are sentenced to jail terms ranging from zero a maximum term set by the user. When a jail term expires, agents leave jail with the same level of grievance as they entered. A plot of some 20,000 iterations of the model, generated a time series of total rebels displaying a hall mark of complex systems: punctuated equilibrium. "Long periods of relative stability are punctuated by outbursts of rebellious activity. And indeed, many major revolutions (e.g., East German) are episodic in fact" (p. 7245).

Epstein's Civil Violence Model influenced subsequent agent-based modeling of insurgencies (e.g., Yiu et al. 2003; Appleget et al. 2014; see overview by Lemos et al. 2013). Uri Wilensky (2004) adapted it in the NetLogo Rebellion Model. Lemos et al. (2014) added complexity to it with a time delay for imprisonment due to fights between rebels and cops, media agents seeking to record those fights, and a feedback mechanism for decreased legitimacy as consequence of number of arrests and media coverage. Those parameters "produced large intermittent peaks of rebellion, which lasted longer and had a more complicated fine structure than those obtained with Epstein's model" (p. 7). The two-actor Civil Violence Model can be extended by including additional types of agents, more elaborate behavioral rules, and less random movement.

Iruba: An ABM of the Guerrilla War Process In one of the earliest efforts to model guerrilla war as a whole, Jim Doran developed Iruba in part to test Guevara's foco theory that "even a very small, dedicated group of insurgents will succeed provided that they have a political as well as military strategy, and provided that there is a significant level of initial support in the population at large" (Doran 2005: 4). His ABM is structured as a network of 32 regions on an island which vary in terrain and uncommitted general populations that provide recruitment pools for both regime forces and insurgents. A successful attack on a regime base may lead to capture of weapons, increased population support of the insurgency, and recruitment to the guerrilla force. As an insurgency grows, regime forces move to regions of strongest insurgent activity. The regime may take "all out" suppression measures. Insurgents may move continually and incautiously from region to region in "hyper mobile" mode, akin to the "flying columns" of the Irish Republican Army, even launch an "all out" attack across the island, like the Viet Cong's Tet offensive during the Vietnam War. "Victory in this model is a matter either of insurgent annihilation, or of the insurgents achieving numerical superiority and hence, by assumption, political power" (p. 3).

Experimental trials of the Iruba model showed that outcomes depended on initial guerrilla band size. Success means that the total insurgent force grows to more than 100,000. When the initial band has fewer than 40 members, insurgent victory occurs in less than 30% of the model runs; a band of 55 guerrillas achieves success 90% of the time. Doran argued that, given the Cuban Revolution started with a band just 21 survivors of the landing party, the Iruba ABM indicates the unreliability of Guevara's foco theory.

> Taken together these results suggest that sufficient preconditions for insurgent success in the Iruba model as calibrated are: a sufficiently large initial band, at least minimal mobility, attack efficiency, some initial population support, and communication processes by which insurgent successes impact the population at large and increase awareness and support for the insurgents. (Doran 2005: 5)

An "all out" regime counter-attack can thwart an insurgency if deployed well before the insurgency force reaches 30,000 men. Doran discussed the limitations of validating the Iruba ABM's wide range of parameters as well as an apparent positive feedback loop: increased insurgent numbers increases population support, which further increases recruitment.

Infantry Warrior Simulation For more than a century, models of combat attrition have successfully applied Lanchester equations to estimate rates of losses for opposing armies. Frederick Lanchester (1916) devised a set of differential equations resulting in three laws of attrition The Square Law applies to model modern warfare with long-range weapons such as artillery and rockets. The side that can concentrate superior numbers on the battlefield can subject the inferior force the heavier fire than it can return. Seymour Deitchman (1962) modified the Lanchester equations to model guerrilla warfare in which an ambushing insurgent force uses directed fire, leading to attrition of the ambushed soldiers consistent with the Square Law. Uncertainties about the numerical values of the attrition coefficients limit the accuracy of the predicted outcomes of an engagement. More recently, agent-based combat models are seen as supplemental to Lanchester equations (Ormrod and Turnbull 2017).

Vikram Mittal (2023) described the Infantry Warrior Simulation (IWARS), a US Army agent-based model for analyzing ground-based military operations at the squad or platoon level. One application assesses the impact of a new technology that increases the lethality of a single fighter to that of multiple fighters. For example, in the Battle for Mosul, from October 2016 to July 2017, entrenched Islamic State (ISIS) guerrillas fought Iraqi forces trying to eliminate them. The casualties on each side fit Lanchester's Square Law. "Although the Iraqi forces suffered fewer casualties compared with ISIS, it is important to consider their numerical advantage with 10 times as many soldiers" (p. 7). In the later stages, ISIS deployed hobby drones to drop grenades on Iraqi soldiers. To evaluate the effectiveness of that new technology, IWARS ran three scenarios of ISIS attacks on soldiers' traffic control points: (1) a baseline of attacks by suicide bombers; (2) drones dropping grenades; (3) a ground-based robot able to shoot down drones. In all three cases, ISIS casualties did not vary, as all were ultimately eliminated by Irai forces. Weaponized drones increased the Irai casualties considerably, but with counter-drone equipment the casualty count was comparable to the base scenario. A second IWARS case study examined the impact of a more accurate rifle in a war involving both urban and rural combat. The side with the new rifle had "a large advantage on the battlefield, shifting the tide of the battle" (p. 12). Mittal argued that ABMs can potentially provide more accurate attrition rates for use in Lanchester equation predictions.

4 Insurgents & Soldiers Fighting ABM

This agent-based model of insurgents and soldiers fighting to eliminate their enemy incorporates elements from two NetLogo ABMs (Wilensky 1999). Uri Wilensky's "NetLogo Flocking Model" (1998) is used to animate small groups of soldiers on patrol searching for insurgents (although often used to model birds flocking, it can also be applied to human collective actions). Yicheng Shen's "Model of Wheeler" (2022a) is a reconstruction of an ABM described by Scott Wheeler (2005a, b) who did not publish his software. Shen (2022b) provided additional details of his recreated Wheeler model and an improved version. The paragraphs below give an overview of "Insurgents and Soldiers Fighting". This adaptation includes neither soldier reinforcements nor civilian agents who interact with soldiers and insurgents.

The world wraps both horizontally and vertically, which means that agents moving off one side of the world will reappear on the opposite side. The terrain is a stylized jungle with darker green patches indicating higher-density vegetation. No village and town constructions are created. The names of two types of agents, their colors, and movements are:

Insurgents (red) are armed rebels fighting against soldiers of the central government. To setup a model run, the investigator uses a slide to randomly scatter a number of insurgents across the world. To conceal themselves, insurgents move to an area having the highest density of vegetation within radius 30. They remain hidden unless detected by soldiers on patrol. If detected, an insurgent moves away from the detector toward another nearby high-density patch. Insurgents remain detected for 20 ticks. If insurgents spot any unalerted soldiers within radius 8, they attack. But they do not attack alerted soldiers.

Soldiers (blue) are initially placed at the center of the world. They move randomly and sometimes jointly in small patrols for safety, using Wilensky's NetLogo code for agent flocking. Soldiers become alerted if: (1) insurgents are detected within radius 3; (2) any of their patrol mates become alerted; or (3) an insurgent attacks any patrol mate. When alerted, soldiers move toward the threat and try to kill insurgents.

An attack by either side always results in killing, depending on the kill-rates selected by two sliders. Both default rates are set at 50%, meaning soldiers and insurgents have equal chances of killing the enemy or dying by return fire. The outcome is determined by a random number between 1 and 100. For example, when the default rate is 50, a soldier with a random number greater than 50 kills the insurgent, while a random number 50 or lower results in the insurgent killing the soldier. Using the sliders to raise or lower the two kill rates, a researcher can simulate differences in effective weapons technology available to each side, which reflects the Lanchester Square Law that a force possessing more lethal arms will attrit its enemies. A model run ends when one side entirely eliminates its enemy. A plot shows the numbers of surviving insurgents and soldiers over time while two monitors display the percentages of dead combatants.

Figure 1 is a screen capture after 10 clicks of a model run with 15 insurgents and 15 soldiers. Single soldiers and patrols are moving away from the center of the world, while insurgents are fleeing toward high-density patches to avoid detection. Figure 2 shows the results at the end of run after 112 ticks, when all 15 soldiers had been eliminated and 8 of the 15 insurgents were dead.

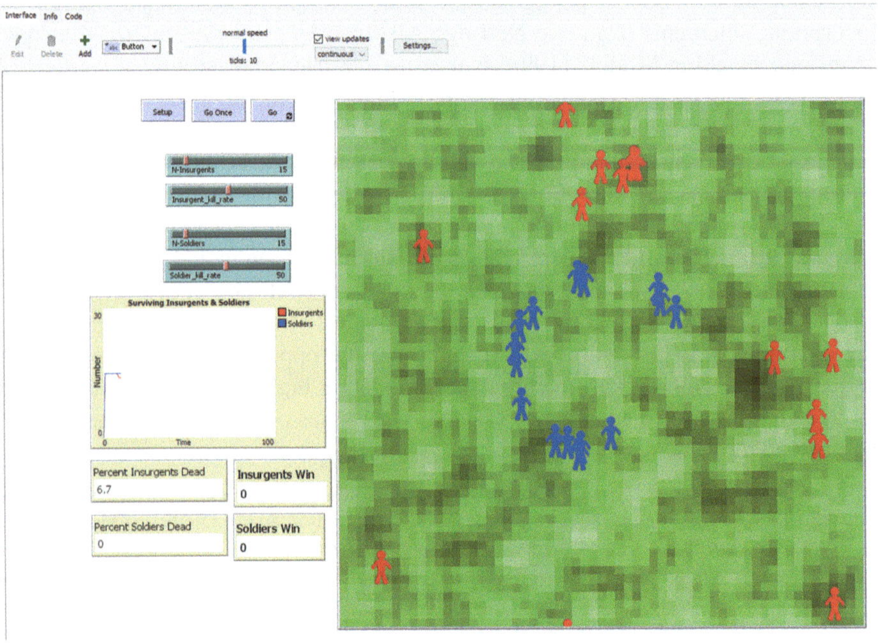

Fig. 1 Model run with 15 insurgents and 15 soldiers after 10 ticks

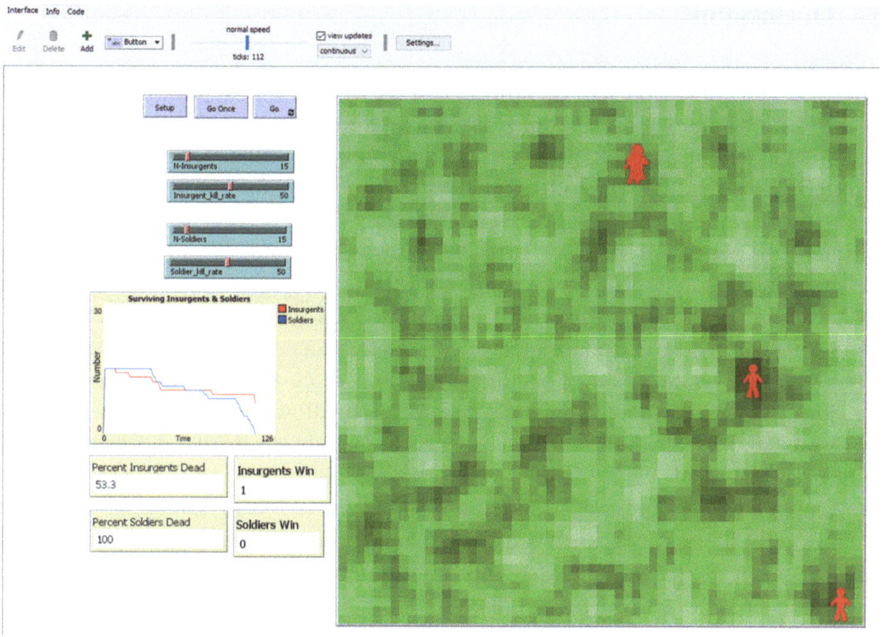

Fig. 2 End of model run with all soldiers eliminated and 8 insurgents dead

Table 1 Experiments with different numbers and kill rates of insurgents and soldiers

	Model 8.1	Model 8.2	Model 8.3	Model 8.4
Number of insurgents	15	15	15	15
Number of soldiers	15	20	20	20
Insurgent kill rate	.50	.50	.60	.70
Soldier kill rate	.50	.50	.50	.50
Percent insurgents killed	84.6%	72.3%	83.4%	64.0%
Percent soldiers killed	87.3%	74.6%	87.9%	98.3%
Percent insurgent wins	53%	25%	55%	88%[a]
Percent soldier wins	47%	75%	45%	9%[a]
Mean ticks	363	316	313	335
Number of experiments	200	200	200	200

[a] In 3% of runs all soldiers and all insurgents were eliminated

Table 1 summarizes the outcomes of four experiments with 200 runs each. The first experiment with 15 soldiers and 15 insurgents used the default kill rates. Insurgents eliminated soldiers slightly more often than the reverse outcome (53% to 47%). The second experiment also used the default kill rates, but increased the number of soldiers to 20. That larger counterinsurgency force was victorious in 75% of the runs compared to 25% insurgent wins. The last two experiments kept the unequal numbers of agents, but increased the insurgents' lethality, with consequence that they were increasingly victorious against the larger number of soldiers.

5 Discussion

The Insurgents and Soldiers Fighting ABM experiments resulted in high percentages of combatant deaths on both sides because constant probabilities of killing in every engagement were set at .50 or higher. A future modification would allow the killing rates to fluctuate from attack to attack. A future extension would add civilian agents to the collective action scenario as Wheeler (2005a, b) did in his Counterinsurgency ABM. Civilian behavior was restricted to informing soldiers about known insurgent threats. A more complex dynamic would allow for injuries and deaths to civilians ("collateral damage") suffered during crossfires between soldiers and insurgents. As a consequence, angry neighbors could shift their allegiances from one side to the other depending on which force they blame most (Bennett 2008; Shen 2022b). Such an enhanced ABM might allow evaluation of the competing Mao and Guevara theories of the importance to win the hearts-and-minds of the populace.

Weaponry and communication technologies evolved vastly since the primitive conditions of the mid-twentieth century, raising the question whether modern insurgencies can be characterized "Mao with smart phones and internet?" (Sink 2020). If insurgents are sponsored by a supplier of cutting-edge technology, they are more likely to succeed against the government. Perhaps the most famous example was the US Central Intelligence Agency providing the Afghan Mujahideen guerrillas with surface-to-air Stinger missiles during the 1979–1989 Soviet-Afghan War. By destroying numerous helicopter gunships, the missiles offset the invader's huge technological superiority, boosted insurgent morale, and forced the eventual Soviet withdrawal, and toppling of the Afghan communist regime (Kuperman 1999; Westermann 1999). A contemporary instance is Iran's underwriting of an extensive network of violent nonstate actors across the Middle East, deemed terrorist organizations by the West: Islamic State in Iraq and Syria, Hezbollah, Hamas, and Houthis. Its drone and missile programs enable the Houthi militants in Yemen to drive global shipping away from the Suez Canal in 2024. Iran even sold thousands of Shahed-136 armed drones to Russia for use in the war against Ukraine (Citrinowicz 2024). Artificial intelligence has already transformed every dimension of warfare from training, surveillance, cyberspace, logistics, autonomous vehicles, to robotic warriors. Will AI lead to a new era of enhanced or irrelevant insurgencies? The final chapter speculates on some possible futures of network collective action.

References

Appleget J, Burks R, Jaye M (2014) A demonstration of ABM validation techniques by applying docking to the Epstein civil violence model. J Def Model Simul 11:403–411

Bennett S (2008) Governments, civilians, and the evolution of insurgency: modeling the early dynamics of insurgencies. J Artif Soc Soc Simul 11(4):7. https://jasss.soc.surrey.ac.uk/11/4/7.html. Accessed 7 July 2024

Boot M (2013) The evolution of irregular war: insurgents and guerrillas from Akkadia to Afghanistan. Foreign Aff 92:100–114

Childs MD (1995) An historical critique of the emergence and evolution of Ernesto Che Guevara's Foco theory. J Lat Am Stud 27:593–624

Chomsky A (2015) A history of the Cuban revolution, 2nd edn. Wiley-Blackwell, New York

Citrinowicz D (2024) Iran is on its way to replacing Russia as a leading arms exporter. The US needs a strategy to counter this trend. Atlantic Council, 2 February. www.atlanticcouncil.org/blogs/iransource/iran-drone-uavs-russia. Accessed 12 Aug 2024

Deitchman SJ (1962) A Lanchester model of guerrilla warfare. Oper Res 10:818–827

Doran J (2005) Iruba: an agent-based model of the guerrilla war process. In: Representing social reality, pre-proceedings of the third conference of the European Social Simulation Association, pp 198–205

Epstein JM (2002) Modeling civil violence: an agent-based computational approach. Proc Natl Acad Sci 99(suppl. 3):7243–7250

Fonay KW (2013) On guerrilla warfare: two takes, Mao vs. Guevara. Small Wars Journal. https://smallwarsjournal.com/jrnl/art/on-guerrilla-warfare-two-takes-mao-vs-guevara. Accessed 3 July 2024

Guevara EC (1960) La guerra de guerrillas. Talleres Tipográficos del INRA. del Departamento de Instrucción del MINFAR

Hobson RB, Moghadam A (2024) Terrorism, guerrilla, and the labeling of militant groups. Terror Polit Violenc 36:567–584

Johnson J (2006) From Cuba to Bolivia: Guevara's Foco theory in practice. Innov J Polit 6:26–32

Knoke D (2013) 'It takes a network': the rise and fall of social network analysis in U.S. Army counterinsurgency doctrine. Connect 33:1–10

Kruijt D (2023) Che Guevara and guerrilla warfare. Globalizations 20:1528–1539

Kuperman AJ (1999) The stinger missile and US intervention in Afghanistan. Polit Sci Quart 114(2):219–263

Lanchester FW (1916) Aircraft in warfare: the dawn of the fourth arm. Constable, London

Lemos C, Coelho H, Lopes RJ (2013) Agent-based modeling of social conflict, civil violence and revolution: state-of-the-art-review and further prospects. In: Proceedings of the 11th European workshop on multi-agent systems. CEUR-WS

Lemos C, Lopes RJ, Coelho H (2014) An agent-based model of civil violence with imprisonment delay and legitimacy feedback. In: Second World Conference on Complex Systems (WCCS), Agadir, Morocco, 10–12 November, pp 524–529. https://ieeexplore.ieee.org/document/7060998. Accessed 17 Jan 2021

Mao Z (1930) A single spark can start a prairie fire. Selected Works of Mao Tse-Tung. www.marxists.org/reference/archive/mao/selected-works/volume-1/mswv1_6.htm. Accessed 9 July 2024

Mao Z (1937) On guerrilla warfare. www.marxists.org/reference/archive/mao/works/1937/guerrilla-warfare. Accessed 5 July 2024

McChrystal SA (2011) It takes a network: the new front line of modern warfare. Foreign Policy, March/April:1–6

Mittal V (2023) Estimating attrition coefficients for the Lanchester equations from small-unit combat models. J Def Model Simul:1–13

Ormrod D, Turnbull B (2017) Attrition rates and maneuver in agent-based simulation models. J Def Model Simul 14:257–272

Pérez-Stable M (2011) The Cuban revolution: origins, course, and legacy, 3rd edn. Oxford University Press, Oxford

Petraeus DH, Amos JF (2006) U.S. Army/Marine counterinsurgency field manual 3-24. Headquarters Department of the Army, Washington, DC

Shen Y (2022a) Model of wheeler. NetLogo Modeling Commons. Northwestern University Center for Connected Learning and Computer-Based Modeling, Evanston. https://modelingcommons.org/account/models/5618. Accessed 3 July 2024

Shen Y (2022b) Be steady and popular: a modern counter-insurgency ABM. NetLogo Modeling Commons. https://modelingcommons.org/account/models/5618. Accessed 3 July 2024
Sink JT (2020) Mao with smart phones and Internet? A comparison of classic guerrilla warfare with fourth and fifth generation warfare using an agent-based model for simulation. Department of Politics and Policy doctoral dissertation. Claremont Graduate University, Claremont
U.S. Joint Chiefs of Staff (2018) Counterinsurgency. Joint Publication, Washington, DC, pp 3–24
Ünal MC (2016) Terrorism versus insurgency: a conceptual analysis. Crime Law Soc Chang 66:21–57
Westermann EB (1999) The limits of soviet airpower: the failure of military coercion in Afghanistan, 1979–89. J Confl Stud 19(2):39–71
Wheeler S (2005a) It pays to be popular: a study of civilian assistance and guerilla warfare. J Artif Soc Soc Simul 8(4). www.jasss.org/8/4/9.html. Accessed 3 July 2024
Wheeler S (2005b) On the suitability of NetLogo for the modelling of civilian assistance and guerrilla warfare. DSTO-TN-0623. Australian Defence Science and Technology Organization, Systems Sciences Laboratory, Edinburgh
Wilensky U (1998) NetLogo flocking model. Center for Connected Learning and Computer-Based Modeling, Northwestern University, Evanston. http://ccl.northwestern.edu/netlogo/models/Flocking. Accessed 16 Jan 2025
Wilensky U (1999) NetLogo. Center for Connected Learning and Computer-Based Modeling. Northwestern University, Evanston. http://ccl.northwestern.edu/netlogo. Accessed 16 Jan 2025
Wilensky U (2004) NetLogo Rebellion Model. Northwestern University Center for Connected Learning and Computer-Based Modeling, Evanston. http://ccl.northwestern.edu/netlogo/models/Rebellion. Accessed 22 May 2022
Yiu SY, Gill AW, Shi P (2003) Using agent based distillations to model civil violence management. J Battlef Technol 6:27–32

Anticipating

Keywords Artificial intelligence collective action · Common pool resources · Economic collective action · Global warming collective action

1 Introduction

The agent-based models explored in this book emphasize the political behaviors of persons. Time and space limitations prevented inclusion of many other forms of collective action. This concluding chapter briefly outlines three additional topics where ABMs might be fruitfully applied: economic collective actions by corporations; global warming collective actions by governments; and artificial intelligent collective actions by nonhuman entities just recently emerging.

2 Economic Collective Action

Economic theories of perfect competitive markets describe numerous buyers and sellers meeting to exchange money for goods or services at a prices determined by supply and demand. The participants have no prior relationships with one another that might affect their transactions. Self-regulating markets do not occur in any real-world economy. Indeed, as Karl Polyani (1944: 3) observed, "Such an institution could not exist for any length of time without annihilating the human and natural substance of society; it would have physically destroyed man and transformed his surroundings into a wilderness." The rational, subjective utility-maximizing *homo economicus* who always acts for selfish benefits ignores the prevalence of empathy, trust, sympathy, reciprocity, compassion, altruism, and cooperation in economic behavior. A more realistic depiction of capitalism must acknowledge that economic activities are embedded in multiplex networks of political, social, cultural, and religious relations that promote or restrain market behaviors (Knoke and Bokun 2020).

D. Knoke, *Network Collective Action*, Lecture Notes in Social Networks,
https://doi.org/10.1007/978-3-031-86199-4_9

The diversity of economic collective actions encompasses labor unionization and collective bargaining, price-fixing collusion, consumer boycotts, cooperatives, and legislative lobbying. This section confines conjectures to strategic alliances among corporations.

A strategic alliance "is a partnership between two or more firms that remain legally independent but share managerial control over the performance of assigned tasks as well as the benefits of the joint tasks" (Mani and Knoke 2011b: 822). Corporations create alliances for many purposes, such as achieving public legitimacy, coping with market uncertainties, innovating new products and services, modifying industry standards, penetrating international markets, and influencing policymaking (Knoke 2012: 132). Because firms often have multiple alliances with different partners, simultaneously and over time, the set of relations constitutes a network of interorganizational ties that can be analyzed with a variety of social network methods to reveal structural patterns affecting economic outcomes. Examples include alliances, joint ventures, and share acquisitions among U.S. firms 1982–2008 (Mani and Knoke 2011a) and alliances among transnational firms in the Global Information Sector 1990–2001 (Granados and Knoke 2013; Todeva et al. 2019). Recent research examined Cleansky, a European multilateral "coopetition" network to develop zero-emission aircraft (Rouyre et al. 2024); the impact of R&D alliance efficiency and formal governance on innovation outcomes in the U.S. biopharmaceutical industry (Wang 2024); and mapping a global firm-level supply chain network to respond to disruptions such as the Covid-19 pandemic (Pichler et al. 2023).

Agent-based models of economic behavior have mostly consisted of individual actors seeking to maximize personal gains (Hamill and Gilbert 2016). Less common were ABMs portraying collective economic actions. Some analysts explored interorganizational collaboration in innovation networks (e.g., Cozzoni et al. 2021; Esmaelnezhad et al. 2023) while other modelers studied supply-chain networks (e.g., Gold et al. 2020; Proselkov et al. 2024). The relative paucity of ABMs analyzing economic collective action is a great opportunity for researchers to develop this emergent field. A primary objective should be to model the evolution of collaborative networks involving follow-on alliances among current or previous partners, the creation of new alliances between partners-of-partners, and alliances among previously unconnected firms. A second focus should examine the consequences of economic collective action, including such performance outcomes as growth, return on investments, profits and losses, market shares, mergers and acquisition, bankruptcies, and dissolutions.

3 Global Warming Collective Action

Planetary overheating is the existential crisis of our time. Efforts to mitigate and reduce its damages cannot rely solely on collective actions by individual citizens. Decarbonization actions of governments and other institutions are urgently required, but are falling woefully short of what is needed. Can agent-based models of collective action on global warming explain the problem and suggest solutions?

Six decades ago, economist Mancur Olson published *The Logic of Collective Action: Public Goods and the Theory of Groups* (1965). Public goods are commodities or services that a governments, private organizations, or individuals provide to all members of a social system. They exhibit non-exclusivity (use by one person does prevent use by others) and non-rivalry (using a good does not reduce its value for others). Governments use taxes to provide such public goods as roads, parks, libraries, schools, and national defense. Private-sector organizations, such as business associations and labor unions, often seek to influence the creation of public goods by lobbying government officials to pass legislation or regulations that benefit their members. Olson argued that rational actors would not contribute sufficient resources to any collective action to produce purely public goods. Instead, people and organizations would take a "free ride" by contributing little or nothing, yet enjoy any resulting public good from which, by definition, they cannot be excluded. A few wealthy individuals might contribute something but the total amount would likely be suboptimal. Olson theorized that organizations would be compelled to offer "selective incentives" whose costs are more than covered by members' dues. Enticements include "magazine subscriptions, group insurance, social gatherings, certification and training programs, and similar benefits from which nonmembers could be effectively excluded" (Knoke 2006: 74). Olson's free-rider hypothesis explained that lobbying groups are a by-product of interest associations that monopolize private goods markets. Summarizing decades of research on Olson's collective action propositions, Todd Sandler (2015: 215) concluded that "they remain valid in many essential stylized cases … that represent numerous real-world situations. His provocative maxims motivated 50 years of research, experiments, and empirical studies that have sharpened everyone's understanding of collective action."

Common-pool resources (CPR) are a type of public good susceptible to destructive overuse. Alternatively described as "the commons", they include pastures, ranges, forests, lakes, rivers, ground waters that local community members have access for hunting, fishing, sheep and cattle grazing, logging, and other resource-extractive activities. In contrast to pure public goods, rivalry in a CPR results in each user's gain subtracting from the resources available to others, the "tragedy of the commons" (Hardin 1968). Efforts to regulate oceanic fish stocks failed to prevent 90 percent of global fish stocks from becoming fully- or over-exploited (Hollway 2015: 4). Similarly, attempts to halt global warming by reducing greenhouse gas emissions have foundered in part because of ineffective mechanisms for monitoring and sanctioning nations failing to fulfill their contributions to their pledged goals (Knoke 2023).

Elinor Ostrom challenged the conventional wisdom that conflict between individual rationality and collective action inevitably results in suboptimal CPR benefits. Agent-based modeling could complement empirical field research and laboratory experiments to understand and improve the governance of social-ecological systems (Janssen and Ostrom 2006). Ostrom proposed a polycentric governance theory in which many self-organizing efforts generate local and regional communication, trust-building, policy experimentation, and learning (Ostrom 2009, 2010a, b). Core elements of polycentric collective action include:

> multiple governing authorities at different scales rather than a monocentric unit … Each unit within a polycentric system exercises considerable independence to make norms and rules within a specific domain (such as a family, a firm, a local government, a network of local governments, a state or province, a region, a national government, or an international regime) (Ostrom 2010a: 552).

Ostrom's modified collective action model asserts that people and organizations will collaborate as trust and confidence increase through working together to create a CPR public good. They become more willing to undertake necessary collective actions that increase their short-term costs because they believe greater longer-term benefits will accrue to themselves and others, and they believe that most other participants are also cooperating even as planet Earth inexorably heats. Evidence is mixed that polycentric collective action on global warming is occurring. Although multiple governing authorities are operating on many different scales, "the jury is still out on how *capable* it is of significantly accelerating decarbonization" (Jordan et al. 2018: 363). Resolving the crisis before time runs out will depend on many factors aligning precisely.

4 Artificial Intelligence Collective Action

In late 2022, OpenAI launched ChatGPT4, a generative artificial intelligence algorithm, which within months became the best-selling software in history. It also provoked an open letter, signed by thousands of computer scientist and tech leaders, calling for a pause for at least 6 months pause on advanced AI development because of "profound risks to society and humanity" (Metz and Schmidt 2023). The Center for AI Safety, a nonprofit association, released a one-sentence statement signed by more than 350 industry leaders: "Mitigating the risk of extinction from A.I. should be a global priority alongside other societal-scale risks, such as pandemics and nuclear war" (Roose 2023). At a Senate hearing, the CEO of OpenAI urged tighter regulation of AI for its potential harms. Despite all the hoopla and fear-mongering, the world is very far from creating the self-aware androids, terminators, and killer robots envisioned by science fiction authors from Isaac Asimov (1950) to Martha Wells (2020).

Generative artificial intelligence (GAI) uses large language models (LLMs) to create textual, photographic, video, and verbal communications based on a few specific prompts from humans. GAI are trained, using unsupervised or semi-supervised learning, on vast quantities of documents, images, and sounds, typically scraped from the World Wide Web. For example, social science researchers use GAI to content-code texts and images, while students generate essays and course papers (Davidson 2024). Infringing on proprietary and copyrighted materials raises knotty problems of privacy, legality, and ethics (e.g., Krotov and Johnson 2023; Wang et al.

2024). GAI can also detect malicious traffic and strengthen cybersecurity (Malatji and Tolah 2024).

A major challenge for agent-based modelers is how to integrate artificial intelligence and agents so that the agent's behavioral rules are enhanced or even replaced by AI. The "mechanisms can range from simple pre-defined decisions to well-designed reward functions through which the agent learns to optimize its decisions and achieve the best possible performance" (Hauff and Lurz 2022: 1). The goal is to replace the rigid rules constraining agent decisions with "hybrid" models consisting of autonomous, self-learning agents created by machine learning and expert interactions (Giabbanelli et al. 2017; Ionescu et al. 2024). Ultimately, agents will independently find and apply new behavioral rules without the intervention of human modelers. AI-ABMs synergies would create more accurate and credible models of complex human behaviors than the crude modeler-produced ABMs of today. An intriguing video demonstrated how GPT-4 could be used to write NetLogo code incrementally by interpreting simple user instructions (Germinotion Consulting 2023). The implication is that social science modelers will not need extensive coding expertise to develop sophisticated ABMs. The potential risk is that modelers will obtain outcomes without understanding the underlying dynamics that generated them.

5 Not a Conclusion but a Continuation

The human species engaged in collective actions since the hunts that eventually exterminated the woolly mammoth and saber-toothed tiger. Now we are engaged in collective actions that threaten survival our species. We have the capacity to learn from our mistakes. By modeling alternative scenarios with diverse consequences, we can acquire the knowledge and enlightenment to avoid catastrophic tragedy and achieve a more sustainable future.

References

Asimov I (1950) I, robot. Gnome Press, New York

Cozzoni E, Passavanti C, Ponsiglione C, Primario S, Rippa P (2021) Interorganizational collaboration in innovation networks: an agent based model for responsible research and innovation in additive manufacturing. Sustain For 13:7460

Davidson T (2024) Start generating: harnessing generative artificial intelligence for sociological research. Socius 10:23780231241259651

Esmaelnezhad D, Taghizadeh-Yazdi M, Mahdiraji HA, Vrontis D (2023) International strategic alliances for collaborative product innovation: an agent-based scenario analysis in biopharmaceutical industry. J Bus Res 158:113663

Germinotion Consulting (2023) Building agent-based models with GPT-4 and NetLogo. www.youtube.com/watch?v=Gsmq1uuPysM. Accessed 5 Nov 2024

Giabbanelli PJ, Gray SA, Aminpour P (2017) Combining fuzzy cognitive maps with agent-based modeling: frameworks and pitfalls of a powerful hybrid modeling approach to understand human-environment interactions. Environ Model Softw 95:320–325

Gold S, Chesney T, Gruchmann T, Trautrims A (2020) Diffusion of labor standards through supplier-subcontractor networks: an agent-based model. J Ind Ecol 24:1274–1286

Granados F, Knoke D (2013) Organizational status growth and structure: an alliance network analysis. Soc Networks 35:62–74

Hamill L, Gilbert N (2016) Agent-based modelling in economics. Wiley, New York

Hardin G (1968) The tragedy of the commons. Science 162:1243–1248

Hauff M, Lurz A (2022) Agent-based models using artificial intelligence: a literature review. In: Pacific Asia conference on information systems, pp 1–17

Hollway J (2015) The evolution of global fisheries governance, 1960–2010. Doctoral dissertation. Oxford University, Oxford

Ionescu Ş, Delcea C, Chiriţă N, Nica I (2024) Exploring the use of artificial intelligence in agent-based modeling applications: a bibliometric study. Algorithms 17:1–38

Janssen MA, Ostrom E (2006) Governing social-ecological systems. In: Tesfatsion L, Judd KL (eds) Handbook of computational economics, vol 2. North-Holland, Amsterdam, pp 1465–1509

Jordan A, Huitema D, van Asselt H, Forster J (eds) (2018) Governing climate change: the promise and limits of polycentric governance. In: Jordan A et al (eds) Governing climate change: polycentricity in action? Cambridge University Press, New York, pp 359–383

Knoke D (2006) Collective action. In: Becker J, Zafirovski M (eds) The international encyclopedia of economic sociology. Routledge, London, pp 73–78

Knoke D (2012) Economic networks. Polity Press, Cambridge

Knoke D (2023) The economics and economic sociology of collective action on global warming. In: Zafirovski M (ed) The Routledge international handbook of economic sociology. Routledge, London, pp 525–541

Knoke D, Bokun A (2020) Social relations and economy: the effects of social ties and networks on economic behavior. In: Zafirovski M (ed) A modern guide to economic sociology. Edward Elgar Publishing, Cheltenham/Camberley, pp 70–89

Krotov V, Johnson L (2023) Big web data: challenges related to data, technology, legality, and ethics. Bus Horiz 66:481–491

Malatji M, Tolah A (2024) Artificial Intelligence (AI) cybersecurity dimensions: a comprehensive framework for understanding adversarial and offensive AI. AI Ethics:1–28

Mani D, Knoke D (2011a) On intersecting ground: the changing structure of US corporate networks. Soc Netw Anal Min 1:43–58

Mani D, Knoke D (2011b) Strategic Alliance networks. In: Barnett G (ed) Encyclopedia of social networks. Sage, Thousand Oaks, pp 822–825

Metz C, Schmidt G (2023) Elon Musk and others call for pause on A.I., citing 'profound risks to society'. New York Times, 29 March

Olson M (1965) The logic of collective action: public goods and the theory of groups. Harvard University Press, Cambridge, MA

Ostrom E (2009) Polycentric systems as one approach to solving collective-action problems. Edward Elgar

Ostrom E (2010a) Beyond markets and states: polycentric governance of complex economic systems. Am Econ Rev 100:641–672

Ostrom E (2010b) Polycentric systems for coping with collective action and global environmental change. Glob Environ Chang 20:550–557

Pichler A, Diem C, Brintrup A, Lafond F, Magerman G, Buiten G, Choi TY, Carvalho VM, Doyne Farmer J, Thurner S (2023) Building an alliance to map global supply networks. Science 382(6668):270–272

Polyani K (1944) The great transformation. Farrar & Rinehart, New York

Proselkov Y, Zhang J, Liming X, Hofmann E, Choi TY, Rogers D, Brintrup A (2024) Financial ripple effect in complex adaptive supply networks: an agent-based model. Int J Prod Res 62:823–845

Roose K (2023) A.I. poses 'risk of extinction,' industry leaders warn. New York Times, 30 May

Rouyre A, Fernandez A-S, Bruyaka O (2024) Big problems require large collective actions: managing multilateral coopetition in strategic innovation networks. Technovation 132:1–9

Sandler T (2015) Collective action: fifty years later. Public Choice 164(3–4):195–216

Todeva E, Knoke D, Keskinova D (2019) Multi-stage clustering with complementary structural analysis of 2-mode networks. In: Proceedings of the 2019 IEEE/ACM international conference on advances in social networks analysis and mining, pp 771–778

Wang T (2024) R&D alliance and innovation: the interaction of network structure and the quality of the relationship. Br J Manag 35:1–17

Wang Z, Chen C, Sehwag V, Pan M, Lyu L (2024) Evaluating and mitigating IP infringement in visual generative AI. arXiv preprint arXiv:2406.04662

Wells M (2020) Network effect: a Murderbot novel. Tor Books, New York

Correction to: Network Collective Action

Correction to: D. Knoke, *Network Collective Action*, Lecture Notes in Social Networks, https://doi.org/10.1007/978-3-031-86199-4

The original version of this book is published without the 'About the Book' and ExtraSupplementaryMaterials for chapters 3,4,5,6,7,8 that is to be viewed on Springer Link at the chapter level.

The updated version of this Book can be found at https://doi.org/10.1007/978-3-031-86199-4

D. Knoke, *Network Collective Action*, Lecture Notes in Social Networks, https://doi.org/10.1007/978-3-031-86199-4_10

MIX
Papier aus verantwortungsvollen Quellen
Paper from responsible sources
FSC® C105338

If you have any concerns about our products,
you can contact us on
ProductSafety@springernature.com

In case Publisher is established outside the EU,
the EU authorized representative is:
Springer Nature Customer Service Center GmbH
Europaplatz 3, 69115 Heidelberg, Germany

Printed by Libri Plureos GmbH
in Hamburg, Germany